The End of Memory

THE END OF MEMORY

A Natural History of Aging and Alzheimer's

JAY INGRAM

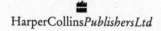
HarperCollins*PublishersLtd*

Published by HarperCollins Publishers Ltd

First edition

HarperCollins books may be purchased for educational, business,
or sales promotional use through our Special Markets Department.

HarperCollins Publishers Ltd
2 Bloor Street East, 20th Floor
Toronto, Ontario, Canada
M4W 1A8

www.harpercollins.ca

Library and Archives Canada Cataloguing in Publication
information is available upon request

ISBN 978-1-44343-576-5

Printed and bound in the United States of America
RRD 9 8 7 6 5 4 3 2 1

To my father, Ralph Ingram,
for demonstrating dignity, patience and love
in caring for my mother through
her years of dementia

Contents

Introduction

I forget exactly what I was looking for when I came across an editorial in the journal *Neurology* titled, "Mom and Me." It referenced a really cool piece of research showing that people whose mothers had had Alzheimer's disease could exhibit the same disruptions of brain metabolism as patients with Alzheimer's and yet be cognitively intact. That is, in the absence of symptoms of any kind, their brains nonetheless seemed to be on the way to Alzheimer's. At the same time, the brains of children of *fathers* with Alzheimer's were ticking along just fine. This convinced the researchers that because the mitochondria (the so-called power plants of the cell) are inherited through the maternal line, Alzheimer's is a disease of the mitochondria. It was good science, but it was more than that.

Those data hit home for me because at the end of her life, my mother was bedridden and unaware. She was said to have Alzheimer's, but since no autopsy was done, that was pretty much a guess. A guess with the odds on its side, true, but not a diagnosis. Not that it would really have mattered. With the exception of some drugs that slow the process for a year, more or less, there is no way of delaying the cognitive decline that *is* dementia. And now it seemed as if I was running the risk of getting the same thing.

I've had what I think is a pretty typical exposure to Alzheimer's. I learned most from helping take care of my favourite aunt as she dwindled away. My aunt, my mother's sister, visited all the familiar checkpoints: forgetting to eat, not being able to use the pill minder because she didn't know *today* was Tuesday, wandering when first moved into a home because we had waited too long to move her. But she would have protested loudly—refused actually—if we had tried to move her any earlier. Eventually, even her good humour deserted her. My father-in-law declined a little faster but in roughly the same way.

But I didn't write this book because I've had family members die of dementia (likely Alzheimer's). Most people have had some sort of experience with the disease—and many have been eloquent about the experience, some of them giving first-person accounts of what it's actually like, others telling their stories from a caregiver's or family member's point of view. As I began to think more about Alzheimer's, I wanted an anatomy of the disease, a natural history. Not a guide to caregiving or diet recommendations or a description of an individual's experience. But a scientific account: Where does it come from? What causes it? Is it a natural part of aging? How are we trying to combat it?

The science of Alzheimer's disease is complex and extremely challenging. As fascinating as any medical mystery, it is unique among them. The emotionally draining personal experience of the disease and the very real threat to health care systems as the numbers of Alzheimer's patients worldwide accelerates have combined to place enormous pressure on Alzheimer's science to come up with a treatment. Alzheimer's is, after all, "the plague of the twenty-first century."

But, of course, it isn't just of the twenty-first century. Long before Alois Alzheimer's name was attached to the disease, the medical world was aware of dementia and described it in terms that are immediately familiar to us. It was first called "Alzheimer's disease" about a hundred years ago, and while that coinage created a brief flurry of interest at the time, it wasn't until the mid-1970s that it became recognized as a disease, rather than a common companion of old age. Since that time, we've been living in a nearly unprecedented era of concentration on a single illness. Or at least it felt that way to me until I saw the stats. In the U.S., the National Institutes of Health (NIH) spends over $6 billion a year on cancer research, $4 billion on heart disease and $3 billion on HIV/AIDS. Alzheimer's? Just $480 million—nothing compared to the cost of care, which is rising inexorably.[1]

So where do we stand? What exactly is the nature of the beast? That's what this book is about.

The first chapter takes a step back to give a sense of how people have thought about aging and death in the past. Sinning was a recurrent theme; dementia was the punishment. On the other hand, your honest striving for salvation could earn you a long and vibrant life and a virtual lock on a place in heaven. Today, we dread Alzheimer's, but in the past, it was usually considered to be a normal part of aging for some. Chapters 2 and 3 tell the story of the beginnings of Alzheimer's disease, from Auguste Deter, the first patient, through some quiet decades until the 1970s, when Alzheimer's was finally recognized as a worldwide threat.

Chapter 4 takes an extremely rare look at dementia from both the outside and the inside; in both cases, the "patient" is Jonathan Swift. In Chapter 5, you'll meet, likely for the first time, Abraham

Trembley, a genius scientist who grabbed everyone's attention with his "immortal" hydra and helped kick-start studies of the biology of aging. Then, in Chapter 6, a closer look at one amazing phenomenon, the inexorable increase in life expectancy over the last 175 years, and James Fries' theory of the "compression of morbidity."

Chapters 7 and 8 are paired. The first is a look at what happens as the brain ages naturally—not with dementia, just healthy aging (assuming these processes are actually different). In Chapter 8, we return to Alzheimer's lab and look over his shoulder to see what he saw in Auguste Deter's brain—the crucial differences that set it apart from healthy aging brains, the differences that are still the basis of an Alzheimer's diagnosis (called "plaques" and "tangles"). They dominated Alzheimer's description of what he saw on his slides, and they dominate thinking about the disease today.

I introduce the Nun Study in Chapter 9, partly because it is one of my favourites (a brilliant idea for a long-term study) and partly because its results underline the fact that Alzheimer's is a complex disease. What might look simple at first glance (if there are plaques in the brain, there is Alzheimer's) turns out not to be (many completely healthy, whip-smart elders have brains absolutely ridden with plaques). The most astonishing result from the Nun Study is that essays written by young novitiates in their early twenties predicted with surprising accuracy who among them would get Alzheimer's sixty years later. The Nun Study has added enormously to our understanding of the disease.

The significance of the Nun Study is made crystal clear by contrasting the apparently inexorable spread of Alzheimer's in the brain, as I do in Chapter 10, with the resistance to damage that has been labelled "brain reserve" (described in Chapter 11). Much is known about where neurons begin to break down in the brain and how plaques and tangles apparently conspire to spread in all directions from those initial sites. But brain autopsies conducted as part of the Nun Study revealed that many cognitively intact nuns had plaques and tangles in their brains when they died. This observation, combined with those from other studies, led to the brain reserve concept—a mysterious something that protects some individuals from dementia. It turns out there's a long list of factors that might be part of brain reserve, a list that will likely grow over time. Education is one of the most important, and education is thought to be an important player in the small number of studies now emerging which suggest that in some places, especially Europe, the incidence of dementia might be decreasing. This is the subject of Chapter 12; it is a surprising and important, though early, trend that bears watching.

With the shadow of Alzheimer's over all of us, we want to know our chances of getting it and what sorts of treatments will be available if we do. That is the subject of Chapter 13. In a way, the study of the genetics of Alzheimer's is in its infancy, but it's already clear that some genes are important for both early- and late-onset Alzheimer's, and more are being announced all the time. In the long run, the hope is that some of these will lead to preventive treatments. Sadly, at the moment, none of those exist. However, as I describe in Chapter 14, there is a rush of clinical trials underway at the moment, most of which—so far anyway—have come up empty. But to an optimist, every failed trial at least yields information you didn't have before. That's the way it works.

Chapters 15 through 18 depart somewhat from the mainstream by taking four individual features of the disease and exploring each one. There are two female Alzheimer's patients for every male, and while superior female longevity accounts for much of this, other, not-well-characterized influences are in the picture, connected, perhaps, to differences in the male and female brains. One of them might be the female hormone estrogen, but again, its role isn't perfectly clear. Once thought to be the key to maintaining cognitive health through menopause and beyond, experts now generally believe that its beneficial actions are much more limited.

Many will remember the aluminum scare, the idea popular in the 1980s and 1990s that aluminum was the cause of Alzheimer's and all aluminum kitchenware should be thrown out immediately. A substantial body of science lay behind the initial worries about aluminum, but inconsistent research results eventually turned most scientists away. All the same, that trend is a lovely example of how a scientific idea can rise—and then fall.

Alzheimer's is by no means the only form of dementia, although it probably represents 75 per cent of the total. The majority are subtle variations on a theme, with rogue proteins (eerily similar to the infectious prions of mad cow disease) accumulating in different parts of the brain and playing a central role in many of those parts. But the most puzzling—and therefore the most scientifically attractive—is the mysterious dementia found on the island of Guam. Is it caused by dietary toxins? Are those toxins delivered in extreme doses by the consumption of bats? This too is ongoing research.

Chapter 18 focuses on where you live and what you eat. It is simply impossible, and would be very tedious, to review the support for every single claim on behalf of a "dementia-protective" food. There are just too many, and the evidence is scattered and scant. But two are worth examining: one is the assertion that turmeric is responsible for the extremely low rates of Alzheimer's found in India; the other, the incontrovertible evidence that large amounts of sugar in the diet are not good. Today, this evidence is considered so central to the problem of Alzheimer's that some are calling the disease type 3 diabetes.

I've tried to describe the state of our knowledge of Alzheimer's as broadly and evenly as I could. I'm sure there will be researchers who won't like my emphasis on this or that or the fact that I didn't write about their approach to the disease. I'm also confident that those who advocate a particular set of vitamin supplements or dietary guidelines will find my omission of them scandalous. My hope is that when you read this book, you'll gain a much broader and deeper understanding of this disease, which preoccupies us so much. Knowing more might even help in that most difficult of tasks: caring for those who are struggling with the affliction.

And by the way, shortly after I read "Mom and Me," I was fortunate enough to come across another article, this one titled "Exceptional Parental Longevity Associated with Lower Risk of Alzheimer's Disease and Memory Decline."[2] Well, that was a pick-me-up! My mother, demented as she was, lived to ninety-four and my dad, a month shy of ninety-eight. In this study, anything over the age of eighty-five was considered exceptional longevity, putting me in a somewhat safer zone. But it would be so simplistic to think that my risk of becoming demented is slightly higher because of my mother and lower because of my mother *and* my father because those are only two risk factors out of dozens if not hundreds. My experience illustrates a point: in the twenty-first century, we face aging in a way that has never happened before—one eye on the clock and the other on Alzheimer's.

CHAPTER ONE

Facing, or Fearing, Aging

Y ou cannot think about aging today without the shadow of
Alzheimer's disease intruding. Many of us—most of us—are
afraid of it: James Watson, co-discoverer of the structure of
DNA, had a set of Alzheimer's gene locations redacted from his
genome because he didn't want to know whether he was prone to
the disease. He was seventy-nine at the time.

The amounts of money being spent on combating the disease
or caring for patients suffering from it are already astronomical, but
overwhelming increases threaten our future. Even with extensive
global research going on, Alzheimer's is still mysterious and complex.
So dependable treatment, let alone a cure, might be distant.

These are things we already know, and this is what aging in
the twenty-first century is all about, particularly in the Western
world. But in the "time before Alzheimer's"—before dementia
became an issue for all of us—what people thought about aging
was very different. Beginning hundreds of years ago, thoughts and
proclamations about sin, vitality, God's will and the stages of life
all fought for the public's attention. A complicated mix, but the

one main difference between then and now is this: when people in centuries past grappled with the inevitability of aging and death, religion was the place to turn. Religion has a weaker hold on us now, but we still place faith, or at least hope, in medical science. We want research to allow us to enjoy a happy and extended life.

But it hasn't been like that for very long. Once you get a glimpse of how different the experience of aging was centuries ago, it's easier to stand back and assess how overwhelming Alzheimer's is. The disease demands an outsider's vantage point, and oddly enough, it is our ancestors who can furnish this persepctive.

In the fourteenth and fifteenth centuries, most people didn't even know exactly how old they were but measured life, if they did at all, in terms of ages or stages. These might be four (childhood, youth, maturity and old age, sometimes linked to the four seasons) or seven, made famous by the lines in Shakespeare's *As You Like It*:

All the world's a stage,
And all the men and women merely players;
They have their exits and their entrances,
And one man in his time plays many parts,
His acts being seven ages.

The seven stages originated centuries before with the astronomer Ptolemy, in his astrological work, *Tetrabiblos*, based on the influences of the sun, the moon and the five known planets.[1] The Ptolemaic influences were precise: the moon was responsible for guiding the first four years of life, Mercury the next ten, Venus the following eight, the Sun the nineteen years of "young manhood" and Mars, Jupiter and finally Saturn the later years. A planet's qualities were evident in their influence: the moon, so changeable as seen from earth, governs the first four years of life, when brain and body develop dramatically. But near the end of a person's years, the slow-moving Saturn presides over the deceleration of life: "the movements both of body and soul are cooled and impeded in their impulses, enjoyments, desires, and speed; for the natural decline supervenes upon life, which has become worn down with age. . . ."[2]

The idea of stages of life dominated thinking about aging for centuries, though the numbers of stages expanded beyond the original four and/or seven. One of the most persistent elaborations, making appearances in one form or another from the fifteen hundreds to the eighteen hundreds, arrayed people on a pyramidal set of stairs, infants on the first step on the left, fifty-year-olds at the top and greater and greater ages descending the stairs on the right. In some versions, centenarians weren't even accorded their own step but lay horizontal beside the last, the ninety-year-old, step on the right. The American printing company Currier and Ives distributed scores of such images as late as the mid-eighteen hundreds.

Variations on the central theme were abundant: for the longest time, the pyramid featured only men; women first appeared as faithful wives and only as themselves in the nineteenth century. Each stage or step on the pyramid was accompanied by the appropriate symbols: the grim reaper holding an hourglass, saplings on the left mirrored by dead trees on the right, an aged cat dozing beside the fire. These were the nineteenth-century equivalent of the popular graphic of human evolution, with our hominid ancestors on the left transitioning to upright-walking *Homo sapiens* on the right.

There were even board games that played on the idea of life as a series of stages. In 1860 American entrepreneur Milton Bradley (whose eponymous company was eventually acquired by Hasbro in 1984) launched *The Checkered Game of Life*, which ran the course from Infancy to Happy Old Age. Only the odd lucky roll allowed players to avoid Ruin or Poverty, but there was no square labelled Death—though notably, there was a risk of Suicide. Bradley sold tens of thousands of copies of his game.

I dug around in my cupboards and found the 2002 version of Bradley's invention. It bears scant resemblance to the original and has none of the darkness of the one produced in 1860: no squares labelled Crime, Idleness, Disgrace or Poverty. Instead, we have Join Health Club, Buy Sport Utility Vehicle and Have Cosmetic Surgery. That's *The Game of Life* today.

Much art has been devoted to the subject of life's stages: among the most important is a set of four huge canvases called *The Voyage of Life* by American painter Thomas Cole.

I first saw these works in the National Gallery in Washington, DC, years ago, long before I had any interest in the subject, but for a few minutes, I was entranced. The four paintings show a trip down a river, beginning with an infant in a boat emerging from a cave and ending with an old man, still in the boat, setting out on the open ocean. It is all religion: a guardian angel accompanies the man throughout his life (though for the most part, without his knowledge); a shiny white castle hovers in the sky; companion angels flit here and there. Exactly what you might expect from a religious, mid-nineteenth-century artist's rendering of "life as a voyage."[3]

It wasn't just Cole, of course: for centuries, religion had been the *only* significant influence on thinking about the passage of life. Yes, people thought of aging as a series of steps or stages, but that was just the calculation and anticipation part. Religion provided motivation: for instance, to counter the view that aging simply draws one further and further from usefulness and closer and closer to death, the Puritans argued that old age actually had an important purpose. It brought one nearer to salvation, something that no forty-year-old could experience. Therefore, there was an incentive, and a powerful one, to live every last day of one's life in a moral way.

Actually, it seemed to be a good idea to get an early start on that moral *modus vivendi*. One widely held belief was that a full, healthy and enjoyable life to the end was possible only through consistent purity of mind and faithfulness to God; those who either died earlier than they should have or, worse, suffered through their last years were seen as the deserving victims of their own immoral lives. They had only themselves to blame, not Providence. So if you were a sinner, old age would inevitably be miserable. Unfortunately, even if you weren't, there were no guarantees.

So the prominence given to angels and heaven in Cole's paintings was no surprise. But there is much more to these canvases: the boat passes through absolutely fantastical landscapes. The carefree youth gliding on calm waters gives way to a troubled middle-aged man deep in prayer as he is tossed about by the waves. There is no doubt that in the end, nature subdues man, but still, the skies toward which the old voyager drifts are heavenly lit.

An abrupt change occurs between the first two and the last two canvases. For the most part, the first and second paintings represent the dreams and hopes of the young (although even in the second canvas, where the youth sails confidently on smooth waters toward a shining castle in the skies ahead, a glance at the extreme right of the painting reveals an upcoming curve in the river, where the waters are choppy, promising a much rougher voyage). The last two canvases are completely grim and dark, the autumn and winter of life, a period that Cole himself described as characterized by trouble.

The Voyage of Life does not even allow for the possibility of choice; the river ensured that there was only one path, a helpless drift toward the sea, albeit watched over by celestial beings. Whether ensured by them or not, the voyage seems to turn out well, with heaven beckoning in the distance.

I have read other accounts by people who were captivated at first sight by *The Voyage of Life*, but it's still not clear to me why this happens. Maybe the paintings force us to address the "threat" of getting old instead of pretending the phenomenon doesn't exist. Or it might be something as pedestrian as the sheer size of the canvases: each is about one and a half metres by two metres (five feet by six feet). Regardless, ever since the four canvases of *The Voyage of Life* were first exhibited publicly in 1840, they have attracted crowds. A decade later, engraved reproductions were hung in homes just as the steps of life had been decades before. Even today, thousands view *The Voyage of Life*, even though the twenty-first-century attitude toward aging has been thoroughly secularized since the paintings were executed about 175 years ago.

Note that these varied visual treatments of life's stages, despite putting wildly different interpretations on what happened to people as the years passed (and how much responsibility they bore for their fate), all bumped up against the same ceiling: no one could escape the inevitability of it all. You climbed the pyramid, then descended step by step. The passenger on Cole's *Voyage* through a landscape apparently unaffected by the presence of humans was propelled by currents and buffeted by the weather; he had no role other than to hang on and pray for his salvation. Even as you moved through *The Checkered Game of Life*, you had to be passive: you might escape the worst or you might not, but you had no control.

The best expression of this helplessness in the face of inevitable old age and death (and the pre-eminent role of religion) was provided by the theologian and preacher Nathanael Emmons. He sermonized in New England for more than sixty years, dying in 1840 at the impressive age of ninety-five. According to his theology, people could influence to some degree whether or not they might ascend to heaven, but above all, they were dependent on God, and God had absolute authority over who died and when. In the timing of death, human behaviour would not influence Him.

Emmons even argued that God could deliberately end a person's life to underline the fact that He was in total control—the vengeful God. This belief led Emmons to argue, surprisingly, that because God therefore ruled over even the laws of nature, it was impossible to know the "natural" life span of a human. That idea allowed for the possibility that human lives, if God weren't tempted to tamper with them, might be much longer than anyone had ever known:

> As we are not perfectly acquainted with the laws of nature, we cannot absolutely determine that any of those who are dead did actually reach the natural bounds of life. We may, however, form some conjecture upon this subject, by the very few instances of those who have lived an [sic] hundred and twenty, or thirty, or forty, or fifty years. . . . Hence we have great reason to conclude that God has most commonly deprived mankind of the residue of their years. And never allowed one in a thousand or one in a million of the human race to reach the bounds of life which nature has set.[4]

Despite Emmons's assertion that many more years of human life might be possible, the fact that God held them in his hand didn't offer much hope of ever obtaining them. And anyway, there had never been much talk of 150-year-olds; immortality had been forfeited in the Garden of Eden. Centenarians, if they were portrayed at all, were virtually comatose. The limits on human life were all too obvious.

By the mid-eighteen hundreds, though, religion's grip on thoughts of aging and death was, at least in some quarters, beginning to loosen. Practitioners of something called "health reform" argued that while God was definitely still in the picture, it made perfect sense for people to live in responsible and healthy ways in order to maximize the years available to them. In fact, living to extreme old age would merely be a return to those fantastic life spans uncritically recorded in the Bible (Methuselah, for example), so this approach was anything but a rebuff to God. People were advised to avoid tobacco, alcohol, coffee and tea (or at least to moderate their indulgence in these things), and they were told not to engage in excessive sex. On the other hand, they were encouraged to bathe and change their clothes more often and to increase their consumption of vegetables.

Many of the health reform movement's recommendations don't sound out of place today, but some of its most enthusiastic champions didn't know when to stop: they foresaw lives of two or three hundred years or more, based, again, on those biblical claims. Indeed, religion was slow to loosen its grip entirely. In his 1857 book, *Laws of Health*, William Alcott wrote: "It is assumed finally that old age must necessarily be wretched. But old age, whenever it is wretched, is made so by sin. Suffering has no necessary connection with old age, any more than with youth or manhood."[5]

In the mishmash of thinking, emoting, rationalizing and sermonizing about death that characterized the nineteenth century, there was one unique, bizarre, not-sure-whether-to-laugh-or-cry strand of thought that emerged in the late eighteen hundreds. Enunciated by none other than the great Canadian-born physician William Osler, the idea had its roots in Anthony Trollope's 1882 novel, *The Fixed Period*. Everywhere described as "dystopian," *The Fixed Period* describes the country Brittanula, which has established a fixed period for a life: sixty-seven years. At that point, people are sent to a place called "the college" in the town of Necropolis, where, within a year, they are euthanized and then cremated. Trollope is said to have derived inspiration for the book from a seventeenth-century play, *The Old Law*, which he had apparently read shortly before writing his novel. But in doing so, he overlooked a much more recent and forcefully argued version of the same idea in George Miller Beard's 1874 publication, *Legal Responsibility in Old Age*.[6]

Trollope was being satirical; Beard, not really. A doctor, Beard based his book on an address he gave to the Medico-Legal Society of the City of New York in March 1873. At the beginning of this talk, he announced he would speak about the impact of aging on the "mental faculties" and indicate whether that impact impaired the elderly to the point where the legal system would be forced to take notice: "The method by which I sought to learn the law of the relation of age to work was to study in detail the biographies of distinguished men and women of every age."[7]

"All the greatest names of history" were included in Beard's analysis. He collated the ages at which each performed his or her most important work: statesmen their legislation, architects their monuments, philosophers their systems. Here are Beard's key conclusions:

- Eighty per cent of the "work of the world" is accomplished before the age of fifty.

- The most productive fifteen-year period is between thirty and forty-five.
- A large number of Beard's subjects lived to be over seventy, but on average their final twenty years, across the board, were unproductive.

Beard, wanting to ensure that his message hit home, attached precious metals and other materials to the decades, beginning with twenty to thirty as brazen, thirty to forty as gold, then silver, iron and tin. The final decade (between seventy and eighty) was "wooden."* To give some heft to his research, Beard pointed out that this principle was universal: horses' best years, according to him, were between eight and fourteen, and hunting dogs were most effective from two to six. Hens hit their egg-laying peak at the age of three, though they might produce eggs for several more years.

Beard also answered some criticisms. Earlier, for instance, it had been thought that the mind was most active between forty and eighty. When asked why that notion had prevailed until he set things right, he claimed not only that people had been excessively reverent toward the aged but also that it takes time for fame to take hold, with the result that we venerate people whose best work was completed decades before. He also held artists responsible: they immortalized, through paintings and busts, men who had been famous but were now way past their prime.

Finally, he came through on his promise to address the issue of whether the legal system should take into account the impact of aging on the mind. His conclusion: the courts should have experts on "cerebro-pathology" on hand to consider the probability that

---

* In fact, graphing his data produces a shape not unlike the steps-of-life pyramid except that the peak is reached sooner and the decline, as a result, is longer.

testimony was being adversely affected by the accumulated damage of excessive age.

George Miller Beard was thirty-four at the time he wrote *Legal Responsibility in Old Age,* so he could be forgiven for a certain short-sightedness. After all, he wasn't proposing anything terribly radical. But he cast doubt on the veracity of his results by never defining the size of his sample, changing numbers each time he quoted them and not publishing the data or calculations leading to his conclusions. Nonetheless, he received a lot of attention; some have argued that his work paved the way for mandatory retirement.

Trollope then wrote the sci-fi version. There these ideas might have languished had not the pre-eminent physician Dr. William Osler chosen "The Fixed Period" as the title of his last address to the Johns Hopkins University Medical School in 1905.[8] Osler was leaving Johns Hopkins for Oxford, and he was making the point that young faculty were good for a medical school, that in fact most people don't contribute much after forty and that maybe they should be forced to vacate the university at the age of sixty. "As it can be maintained that all the great advances have come from men under 40," he claimed, "so the history of the world shows that a very large proportion of evils may be traced to the sexagenarians—nearly all the great mistakes politically and socially, all of the worst poems, most of the bad pictures, a majority of the bad novels, and not a few of the bad sermons and speeches."[9]

Apparently, he was just getting warmed up because a few moments later, he said: "The teacher's life should have three periods—study until 25, investigation until 40, profession until 60, at which age I would have him retired on a double allowance. Whether Anthony Trollope's suggestions of a college and chloroform should be carried out or not I am a little dubious, as my own time is getting so short."[10]

Today, social media would have been all over that comment. We are used to nearly daily examples of people (who should know better) making untimely or thoughtless remarks, but the public response to Osler's joke about chloroform more than a century ago was actually on the same scale: over-the-top newspaper headlines and public pronouncements, especially from those over sixty who felt that their usefulness, at least as they estimated it, was being denigrated. There were even three deaths that seemed suspiciously linked to Osler's comments, including one of a man who apparently chloroformed himself out of this life days after discussing Osler's speech with others and making it clear that as far as he was concerned, the theories should be put into practice.

The funny thing was that apparently no one who was present when Osler made his remarks took him seriously; he was savaged only in the following day's newspaper reports. Yet the damage was done: a new verb, "to oslerize," briefly came into vogue.

The furor Osler created would have been as incomprehensible to Americans a hundred years earlier as it is unremarkable a hundred years later. Where is God? Are not all human lives held in God's hands? Nathanael Emmons would have been red-faced at the impudence of it all. But as sloppy as George Miller Beard's science appears to have been, it was symbolic of the change felt throughout the nineteenth century. Age was now the province of science.

And what could science potentially do? Manipulate nature. Today, more than a century and a half after the health reform movement bragged about extending human life, there is renewed talk of people reaching the age of 150 years. Not because God is out of the picture or back in the picture but because *science*, or what Emmons called "the laws of nature," is now much better understood. Cole's natural landscapes have been changed forever.

In the eyes of the scientific optimists, the passenger in *The Voyage of Life* will no longer be helplessly tossed around by the currents. And while scientific knowledge is far from perfect, it has advanced enough to encourage some experts on aging to argue that we will soon be able to tinker with the human life span.

We'll go into more detail about the biology of aging in Chapter 5, but for now, there are so many different approaches to the science of aging that a clear path to life extension does not, and may never, exist. However, a growing number of credible experts believe that we can expose and clarify the factors that initiate aging in the human body and even extend life as a result. We really *are* in the age of science.

But even as we dream about living longer, we worry. Alzheimer's now dominates the discussion about aging as no other disease has done. A short while ago, this wasn't the case: heart disease, stroke and cancer were considered to be the principal hazards of, or threats to, a long life. Now, when nearly one of every two people over eighty-five is demented, Alzheimer's casts a cold light on the prospects of living long. It's as if Cole's river of life has suddenly opened a new and hazardous side channel, and a growing number of individuals are drifting down it.

As a result, those brave enough to claim that we will one day live much longer are always careful to make the point that at the same time, we will be obliged to ensure that we'll be healthy and *mentally intact* at those great ages. Predictions that the debilitating, chronic disease of Alzheimer's would suddenly come to our attention when the killing diseases, like pneumonia, were eliminated or controlled have come true.[11] After all, pneumonia was called "the old man's friend."

But this is all twenty-first century: had you lived in the nineteenth century, you wouldn't have been very concerned about extending life or about the accompanying risk of a much diminished quality of that extended life.

Alzheimer's disease today affects up to 10 per cent of those over sixty-five and nearly 50 per cent of those over eighty-five.[12] In Canada and the United States, with a total population of nearly 350 million, there are nearly 6 million cases of Alzheimer's disease. In 1800 the *entire population* of the United States (of European descent) was less than that. At the same time, Canada probably had no more than half a million people. And there was no baby boom cohort, no bulge of older people moving through the population. So two hundred years ago, with life expectancy dramatically lower than it is today, with a much smaller population of people over sixty-five and with God, sin and salvation uppermost in people's minds, thoughts of old age didn't allow much room for dementia. But a little over a hundred years ago, things changed.

Chapter Two

"I have, so to say, lost myself"

On November 26, 1901, in Frankfurt, Germany, a young clinician met with a woman who had been admitted to the municipal mental asylum the day before. She was fifty-one. In the months before her arrival, her behaviour had become increasingly bizarre and disordered. She had become paranoid and irrationally jealous of her husband of twenty-eight years, and her memory was declining precipitously. The doctor made careful notes of his interview with the woman, named Auguste Deter:[1]

> *"What is your name?" "Auguste."*
> *"Last name?" "Auguste."*
> *"What is your husband's name?" "Auguste, I think."*
> *"Your husband?" "Ah, my husband."*

She was inconsistent—able to name a pencil, pen, diary and cigar correctly—but further questioning revealed the depth of her confusion.

"What did I show you?" "I don't know, I don't know."
"It's difficult isn't it?" "So anxious, so anxious."

Some of the objects first named correctly were quickly forgotten. When eating cauliflower and pork, she identified them as
spinach. The doctor questioned her further:

"What month is it now?" "The 11th."
"What is the name of the 11th month?" "The last one, if
not the last one."
"Which one?" "I don't know."

The doctor noted that her difficulties extended beyond being
unable to name things correctly. When reading, she repeated the
same line three times. Even though she could identify the individual letters, she seemed not to understand what she was reading and
even stressed the words in an unusual way. Then, out of the blue,
she said:

"Just now a child called, is he there?"

and here and there short phrases that provide a brief window into
an agonizing decline:

"I do not cut myself."
"I have, so to say, lost myself."

Auguste's decline continued unabated. Eventually, her speech
became incomprehensible, and the only sounds she made were
shouting or humming. For the last year of her life, she remained
virtually mute, apathetic and immobile. She died in April 1906,
shortly before her fifty-sixth birthday.

Her case would likely have remained unremarkable, and even unnoticed, had it not been for the tenacity of her doctor. By the time of Auguste's death, he had moved on from Frankfurt to Munich, where he worked at the Royal Psychiatric Clinic at the university. But he had never forgotten her, and when informed of her death, he asked that her brain be sent to him for study. What he found made Auguste D. well known in neurological circles at the time, and the doctor himself famous. He was Alois Alzheimer.

For a man whose name is now attached to one of the best-known and most-feared diseases of the twenty-first century, Alzheimer was anything but a larger-than-life character. Compassionate, yes, and a good mentor, wandering from one co-worker to the next in his lab, offering advice on what they were seeing through the microscope between puffs on his omni-present cigar — the cigar that was often left to burn out on the lab bench as he lost himself in concentration. Various biographers have tried to make something more colourful of this diffident, hard-working, quiet man, but the best anyone has been able to do is to point out that when he was a student in medical school in Würzburg, Germany, he was a member of a fencing club that augmented its thrusts and parries with a full social life, including good German beer (although that is not properly documented). And in 1887, when he was still a student, he was fined for activity that was vaguely defined as "improperly aroused disturbance of the peace in front of the police station."[2]

But Alzheimer was definitely not one of the high-profile, empire-building psychiatrists and neuroscientists of the day like Sigmund Freud, Carl Jung, Emil Kraepelin or Kraepelin's archrival in Prague, Arnold Pick. As one writer put it, Alzheimer was one of those people who truly had fame thrust upon them, although to be fair, he ran a powerful anatomy lab, and several scientists whose names we associate today with important diseases worked there, including both Hans Gerhard Creutzfeldt and Alfons Maria Jakob of Creutzfeldt-Jakob disease (CJD), and Frederic Lewy. (His "Lewy bodies" are aggregates of protein that collect inside brain cells, in both Parkinson's disease and dementia with Lewy bodies.)

But then Auguste Deter entered hospital, Alzheimer became her clinician and his name, if not his science, became widely known. To be accurate, however, not even in Alzheimer's lifetime did his examination of Auguste Deter appear to be the defining moment of his career. She was simply a case in which he was interested, and once he had received her brain, he proceeded to slice it thinly and apply a variety of revolutionary chemical stains that made microscopic structures pop out from the background. These stains had been invented recently, some by his colleague and close friend Franz Nissl, who had also been best man at his wedding.

Alzheimer's examination of Auguste Deter's brain revealed several abnormalities. First, the number of neurons, the brain cells, was dramatically reduced. Among those cells, but outside them, Alzheimer also discovered dark deposits of an unusual composition, and third, using stains containing silver, he was able to discern dark fibrils sitting in the middle of what appeared to be otherwise normal brain cells. Three features, a triad that even today characterizes the disease.

Alzheimer wasn't the first to identify the deposits (which are now called "plaques") or the fibrils ("tangles"). Both had been spotted in brain tissue by other researchers before them. But it was Alzheimer who made the point that they, together with a significantly reduced population of neurons, coexisted in the brain of someone whose premature symptoms of mental decline he had witnessed himself. This observation was enough to move him to report his findings to a wider audience, which, as it turned out, was largely unappreciative.

It was at the November 1906 meeting of the Southwest German Psychiatrists in Tübingen. Some ninety of Alzheimer's colleagues attended, and he reported what he'd seen in Auguste Deter's brain, apparently failing to create even a ripple of interest. (I have seen a claim that the audience was more interested in the report that followed, on compulsive masturbation, but have been unable to verify this.) No questions were asked, and the chairman of the meeting said, "So then, respected colleague Alzheimer, I thank you for your remarks, clearly there is no desire for discussion."[3]

Alzheimer might understandably have felt let down, although it's been pointed out that even he wasn't sure that he'd discovered anything worthy of bearing his name. At most, he thought Auguste Deter had suffered from a strange, early-onset form of senile dementia. Nevertheless, the history making was out of his hands. This single case, along with a handful of related instances in the next few years, prompted Alzheimer's boss, the powerful Emil Kraepelin, to include the cases in the eighth edition of his immensely influential *Handbook of Psychiatry* (a series of texts on which he built a huge reputation) and to call them "Alzheimer's disease." But even Kraepelin himself admitted that the disease, its interpretation and indeed even its importance were "unclear."

Alzheimer's discovery came at a dynamic time in psychiatry and neuroscience, especially in Europe. Sigmund Freud was aggressively—and persuasively—pursuing psychotherapy and the idea that mental illness could be addressed by talk therapy; scientists like Alzheimer were tackling similar issues but much more biologically, trying to connect the dots between what they saw under the microscope with the symptoms a patient had when alive, an approach made possible only by the recent invention of staining techniques that highlighted different features of the brain landscape.

There were bigger ideas in play too, like competing theories as to the nature of the brain itself: Was it one huge interconnected web or were neurons actually individual units, communicating with each other but remaining distinct? It was an exuberant time for brain science, although you'd never know from the generally dour expressions on mustachioed faces in group photos of the time.

Set against this background, Kraepelin's decision to christen Alzheimer's a new disease without really knowing how significant it was has generated plenty of speculation. Some have argued that amid the intensities of research and its accompanying rivalries, Kraepelin wanted to garner every credit he could to elevate his lab over that of his chief rival, Arnold Pick, in Prague. Calling the disease "Alzheimer's" would then simply be a case of one-upmanship, every new disease being a notch in Kraepelin's belt. In fact, Oskar Fischer, one of Pick's colleagues, had just published a paper showing the presence of plaques in the brains of those who had died of dementia, but he did not take the next step of arguing that he had seen a new disease.

It's also been suggested that Kraepelin, a man devoted to the idea that mental illnesses were brain diseases with some sort of organic basis, needed every example he could lay his hands on to keep pace with, or surpass, Sigmund Freud and his psychoanalytic interpretations. The problem was that Freud, using psychoanalysis, was actually having success, while Kraepelin and the twenty pathologists in his lab were not. And there was more than prestige at stake: the technology in Kraepelin's lab was expensive to maintain, and funding shortfalls were chronic; by contrast, psychoanalysis required no labs at all, and Freud and Jung were doing very well financially.

It is also true that the dementia accompanying old age, well known in Alzheimer's time, was generally assumed to be an unremarkable part of aging, not a disease, while the same set of symptoms seen in someone in their fifties was worth noting and differentiating by giving it a name. We will presumably never know what Kraepelin's real reasons were, but the science of the time was moving forward at exceptional speed, and it could well be that he simply wanted to establish the same sort of priority, as scientists would today.

We might never know what Kraepelin's motives were, but we do know that Alzheimer's description of what he'd seen in Auguste Deter's brain was absolutely accurate and that the skill with which he prepared the tissue samples for the microscope impressed colleagues ninety years later. We know these things from more than simply reading the accounts he published in the early twentieth century. In an effort that would rank with the discovery of an upright carafe on a washstand among the ruins of the *Titanic*, late-twentieth-century German researchers actually found not just Alzheimer's clinical notes but also the actual microscope slides of Auguste Deter's brain tissue.

This saga began innocently enough in the early 1990s, when a Japanese scientist argued that Auguste D. and case no. 2, Johann F., were of such interest and scientific value that rediscovering them would give neurologists invaluable insights, even beyond what Alzheimer had described in his technical articles. Despite the complete absence of any evidence whatsoever that they might still exist (in a country that had experienced two world wars), the scientist, Kohshiro Fujisawa, persisted. "I quite agree with you," he said, "that there is left only a faintest hope to rediscover these brains somewhere in Munich. However, I believe in a miracle. I believe because . . . you German people have a world renowned propensity for die Ordentlichkeit (orderliness) und die Pünktlichkeit (punctuality). These brains must have been kept with great care."[4]

He pushed the right buttons because a variety of German researchers with an interest in Alzheimer's disease resolved to address Fujisawa's faith in German science one way or the other. One, Manuel Graeber, was working at the university where Alzheimer had worked in Munich, but so much time had passed that materials as dated as Alzheimer's had largely been discarded. However, some records had been spared from the trash bin for conservation purposes. Only a few weeks after reading Fujisawa's letter, Graeber and others, searching in the basement of the research institute, located Kraepelin's autopsy book, and this led the way to the discovery of both the original notes that Alzheimer wrote about his second case, Johann F., *and* the microscope slides that Alzheimer had prepared of Johann F.'s brain. A stunning discovery—so remarkable that the scientists brought in the Bavarian State Bureau of Criminal Investigation to verify the age of the ink that had been used to label the slides.

But Auguste D. was still the prize (although Johann F. would turn out, upon closer examination, to be interesting indeed). For a few years, nothing at all developed, but then, in December 1995, psychiatrist Konrad Maurer found the Auguste D. file in the university hospital in Frankfurt, where Alzheimer had first met her. Maurer and others had been looking for it for years and eventually discovered it in a cardboard box of decades-old papers. They uncovered a handful of blue files, one of which appeared to be significantly older than the others. One glance at it and Maurer turned to the others and yelled, "This is Auguste D.!"[5] No wonder he was ecstatic—it was the find of a lifetime to see not just Alzheimer's notes in his own hand but also the halting attempts by Auguste D. to write her own name and even pictures of her.

Still missing were the microscope slides portraying the damage to Auguste D.'s brain. Maurer was hoping they might be in

Frankfurt with Alzheimer's clinical notes, but Graeber was counting on the fact that Alzheimer had had Auguste's brain sent to
Kraepelin's lab in Munich. He was right. An amazing 250 slides of
Auguste D.'s brain, each one labelled with her name, were found
in 1997 when researchers checked the autopsy book for transfers
from other cities, including Frankfurt.

So Fujisawa's faith in Germanic attention to detail was fully
repaid, and he was also right in thinking that the discoveries would
turn out to be more than historical curiosities—probably more
right than even he had expected.

Remember that Auguste D.'s was an extreme case: she was
admitted to hospital with signs of dementia when she was only
fifty-one and died five years later. Alzheimer and his colleagues
recognized that this was an extremely early-onset version of
dementia, not the typical kind arising in old age. But that recognition, and the microscopical examination, were about as far
as they could go. In the decades since, of course, much has been
learned about early-onset Alzheimer's, including the discovery
in the mid-1990s of three genes that cause early-onset familial
Alzheimer's disease. "Cause," not "predispose," because these
are dominant genes, meaning that inheriting even one copy guarantees the disease will develop. Even though these genes are
thought to be responsible for no more than 1 per cent of all cases
of Alzheimer's, their early-onset feature is reminiscent of the
kind of disease Auguste D. seems to have had. But is that really
what she was suffering from?

Manuel Graeber and his co-workers did something that
Alzheimer couldn't have conceived of doing—in fact, couldn't
even have *dreamed* of doing. They removed some of the tissue
from Auguste Deter's brain preserved (by Alzheimer himself) on
a microscope slide and performed state-of-the-art genetic testing

on it. They succeeded in showing that there was a mutation in a gene called "presenilin 1," which creates a biochemical domino effect resulting in early-onset Alzheimer's. Hers was a unique mutation—no other human with the same alteration has ever been found. And it's the slightest of changes: a single amino acid mutation out of more than four hundred in a stringy molecule that weaves in and out of the surface membranes of brain cells. But even a single change, if it's at a crucial place, can cause disaster, and that is exactly what happened. We now know that Auguste Deter had always been destined for early-onset Alzheimer's disease.* And of course, when we look to the future, as Manuel Graeber pointed out, "If a brain as old as Auguste D.'s can relinquish its secrets, so can many others."[6]

The reason Alzheimer couldn't have conceived of this genetic testing was that the timing was wrong. While modern genetics does go back to the mid-1860s and the monk Gregor Mendel's pea-plant breeding,† Mendel's work had no impact on science until 1905, when his experiments and their meaning were revived—independently—by a handful of different scientists. That was the beginning of modern genetics and the concept of the gene.

And in 1905 Auguste Deter was already in severe decline. She died before scientists could figure out what genes actually were or what they were made of, let alone what they actually did. There was no genetic testing available to Alzheimer because there really wasn't any science of genetics. It required James Watson and Francis Crick (who determined the structure of DNA), gene sequencing and the Human Genome Project to make possible

---

* Although a recent attempt to replicate this finding failed to find the gene.

† Mendel's work is pretty tedious stuff, certainly not up to the task of intriguing young students in genetics. The most interesting part is the controversial suggestion that statistically speaking, his results were just too good—data fudging maybe?

what Manuel Graeber did. And it will take even more to move ahead. Like Auguste, Johann F., the second patient, has turned out to be tantalizing. He was certainly demented—as recorded in Alzheimer's own examination notes—when he was only fifty-four years old:[7]

Dulled, slightly euphoric, difficulty with comprehension. He repeatedly echoes the question rather than give an answer, and only solves simple calculations after long delay. When tasked with pointing to certain parts of his body, he continually hesitates. Even after a previous discussion about the kneecap, he identifies a key as a kneecap. He does the same when given a matchbox, then strikes it against his kneecap when asked what to do with it. He does the same with a bar of soap. Other requests, such as unlocking a door or washing his hands, he responds to correctly although he does so exceptionally slowly and with great difficulty.

So that's the mental state Johann F. was in. When he died three years later and his brain was thin-sliced and put under the microscope, Alzheimer, in his careful fashion, noted that while there were innumerable plaques of the same kind that he had seen in Auguste Deter's brain, he found no tangles. Not a single representative of one of the three features that he had assumed were so significant in her brain. Had he missed them? No, he hadn't. Modern examination of the slides revealed no tangles either, although it is true that a couple of parts of the brain known to exhibit fibrils early in the disease were not examined. We know today that some cases of Alzheimer's exhibit plaques but no tangles, but they are in the minority and often involve another neurodegenerative process. So Johann F. turns out to have been intriguing as well.

Some preliminary genetic work was done on his brain tissue too, but nothing remarkable emerged. Unlike Auguste, he has not been tested for the presenilin genes. In a case like this, genetics imitates archaeology, where some portion of an archaeological dig is left untouched in anticipation of better future technology and knowledge. Future geneticists might look at Johann F. quite differently than we do today.

It's ironic that Auguste Deter, the index case of Alzheimer's disease, really had not the common version of the disease that now goes under that name but the much rarer, early-onset version. And even Johann F. didn't turn out to be straightforward.

It's ironic too that Dr. Alzheimer's report to the Southwest German Psychiatrists in Tübingen apparently didn't even attract a single question from an audience of ninety. It was a tiny ember of a case at the time, and even after Alzheimer's name was attached to the disease, for a few decades his research didn't become much more than that.

Hard to imagine that there was a time in living memory when you could have gone for weeks or months never hearing the words "Alzheimer's disease." Of course, back in the nineteenth century, there was no illness that went by that name. But was there dementia? Or is the current wave of Alzheimer's and other dementias something new?

CHAPTER THREE

Has Alzheimer's Always Been with Us?

ven though it's easy to document how attitudes toward aging
shifted wildly over the centuries, with elders sometimes being
blamed for their own woes and sometimes being given hope
for salvation (with God never much more than an arm's length
away), it's much more challenging to figure out what people
thought about the declining mind, the condition that we call
"dementia" today. As we saw in Chapter 1, the last of the four canvases in Thomas Cole's *Voyage of Life* reveals an old man, seated
but upright in his boat, praying as an angel guides him toward
heaven. But what are his thoughts? He seems not to be unaware,
but did Cole's audience in the 1840s look at the aged man and
wonder if he was still of sound mind?

There's one very good reason for wondering: Alzheimer's disease today appears to be an epidemic. In just the last few decades,
it has moved from being under the radar to overwhelming it. And
however pressing are the issues of treatment, care and housing
for Alzheimer's patients in our times, they are only going to be
magnified in the decades to come. Is there something special and

unusual about what's happening with the disease today, something about its nature that has made its prevalence skyrocket? Or has it always been around, more or less the same as it is now, but just not as noticeable because fewer people lived to a great age in the past? Clearly, if the disease was nonexistent or disproportionately rare in centuries gone by, we can be suspicious that something indeed *has* changed to accelerate the spread of Alzheimer's. That would be very important. So it's worth looking back—but with caution. Even the relatively recent psychiatry of the nineteenth century has been called "a remote country," where terms like "dementia," "neuron," "plaques" and "tangles" had yet to be defined. And that's just the nineteenth century. Retreating even further into the past brings more and more uncertainty.[1]

Here's one example: ancient medical treatises are couched in language that seems completely foreign to us. It's not just that they use different medical vocabulary; it's that the fundamental beliefs underpinning them are radically different. Hippocrates, for example, analyzed illness through the balance of four bodily humours: black bile, yellow bile, blood and phlegm. Imbalances among them were thought to be responsible for mental problems. Seeing illness through a filter like this makes it clear how difficult it is to interpret any comments about what might seem to be dementia. As we move forward from those ancient times, it's true that things start to become more familiar, but caution is still the guiding word.

How much dementia existed in the past? First, the label "Alzheimer's disease" isn't helpful because as we saw in the last chapter, it wasn't applied to dementia until the early twentieth century. But the word "dementia" is much older, likely coined in the seventeen hundreds or even earlier, and regardless of the name, the symptoms of something awfully like Alzheimer's disease can be recognized in writings from times much earlier than that. Four thousand years ago, the Egyptians recognized that memory loss could accompany old age, although because they thought the mind existed in one's midsection, it's hard to apply much of this observation to our knowledge of dementia today.

Half a millennium before Christ, Pythagoras, inventor of the theorem relating the square of the hypotenuse to the sum of the squares of the other two sides, divided a human life into segments (multiples of seven—he was a numerical man) and remarked that the last stage, old age, brought with it weakness of mind. In that, although he anticipated them by several centuries, his views were no different from those of the medical greats who followed him, Hippocrates and Galen. Both located dementia (or whatever they called it) in the brain, although they allowed for the possibility that it was triggered somewhere else in the body—a not unreasonable idea.

In general, most of the thinking in ancient times seemed to regard weakening of the mind as an inevitable accompaniment of aging.* Only the Roman writer and orator Cicero stands out, arguing as he did that a healthy mind could be maintained well into old age:

> For my part, I know not only the present generation, but their father, also, and their grandfathers. Nor have I any fear of losing my memory by reading tombstones, according to the vulgar superstition. On the contrary, by reading them I renew my memory of those who are dead and gone. Nor, in point of fact, have I ever heard of any old man forgetting where he had hidden his money. They remember everything

---

* Galen took this notion to an extreme by regarding both dementia and aging itself as diseases.

that interests them: when to answer to their bail, business appointments, who owes them money, and to whom they owe it. What about lawyers, pontiffs, augurs, philosophers, when old? What a multitude of things they remember! Old men retain their intellects well enough, if only they keep their minds active and fully employed.[2]

Now what to make of all this? If dementia or a weakening mind, whatever it's called, is thought to be a normal part of aging, medical experts will not pay much attention to it. It's just life. It would have to be something remarkable to attract interest. No interest, nothing written. So absence of written evidence of dementia carries less weight than you might at first think. Back then, the reality of life was that babies were dying at six months of age and old people were becoming demented.

And yet at the same time, because writings from two thousand years ago occasionally do describe symptoms that we can recognize as dementia, we have at least to admit that something sounding very much like Alzheimer's existed then, even if it seems rare when you consider the paucity of writings about the phenomenon. So it's possible that the illness was thought to be ordinary because it was relatively commonplace, but the picture is murky. Only toward the end of the eighteenth century does dementia suddenly attract serious attention.

Two dynamic French physicians set the tone. The first, Philippe Pinel, Napoleon's doctor, revolutionized the treatment of those who were, in the vernacular of the time, "insane." He abhorred the fact that a mix of psychotic, depressed and demented individuals were all packed together in the infamous Bicêtre asylum in Paris. Declaring that insanity was *not* a crime, he had patients unshackled and ordered radically new, humane treatments to be established. At the same time, he opened the door to a more careful and scientific diagnostic approach to mental maladies. For the first time, people with dementia would not be lumped together with the rest; until this happened, it had been impossible to get a true picture of dementia's incidence.

Pinel's student Jean-Étienne Dominique Esquirol was the other half of this duo. What leaps across the centuries is not so much the man's clinical accomplishments as the elegant detail of his writing. If there were any doubts that he was treating people with dementia, likely Alzheimer's disease, they would be dispensed by descriptions like this:

> *Many of those who are in a state of dementia have lost their memory, even of those things which are most intimately connected with their existence. But it is especially the faculty of recalling impressions recently made, which is essentially changed. They possess only the memory of old persons. They forget in a minute, what they have just seen, heard, said or done. It is the memory of things present which is wanting to them, or rather, memory does not betray them because the sensations being very feeble, as well as the perceptions, scarcely a trace is left after them. Many also are irrational, only because the intermediate ideas do not connect those which precede and follow.*[3]

Esquirol is still of another time, as evidenced by his proposed causes of dementia (menstrual disorders, hemorrhoid surgery, political upheavals). And he was not the first to use the word "dementia" (the French *démence* was already in use, though it stood for several different kinds of afflictions that might or might not be curable and could affect any age). However, he stood apart because he described with a modern voice a set of various dementias triggered by different causes, and his crisp differentiation between dementia and other mental disabilities is much quoted: "A man in a state of dementia is deprived of advantages which he formerly enjoyed; he was a rich man who has become poor. The idiot, on the contrary, has always been in a state of want and misery."[4]

He didn't stop at description but added detailed accounts of the dissection of post-mortem brains. Esquirol was thus a pioneer of the science of linking symptoms to brain stuff. But in the end, it's his words that impress the most:

> *Others still, pass days, months and years, seated in the same place, drawn up in bed, or extended upon the ground. This one is constantly writing, but his sentiments have no connection or coherency. Words succeed words; relating sometimes to his former habits and affections. We can sometimes recognize amidst the incoherence and confusion of what they write, a word or phrase which they repeat, and which is the result of memory. . . . One, in an interminable babble, speaks in a loud voice, constantly repeating the same words. Another, with a sort of continued murmur, utters in a very low tone, certain imperfectly articulated sounds; commencing a phrase without being able to finish it.*[5]

Esquirol and Pinel changed the game, and dementia began to attract new attention as something of medical interest, but still a minority interest, as a significant number still clung to the idea that old age inevitably brought with it the curse of dementia: "In the second epoch of old age, which we date from the beginning of the 81st year, the scene of mortal existence closes, after a great length of life, to which, very fortunately, few of the human species survive. The system returns to the imbecility of the first epoch of infancy."[6]

Again, if we want to get an idea of how common Alzheimer's was before the twentieth century, we have to be very careful of what we read and how we interpret it. Esquirol's writings made it crystal clear that dementia was present and accounted for, but in what numbers? The picture is complicated by the fact that while diagnoses in the past were indeed based on the symptoms of a patient, those symptoms were interpreted in the light of the knowledge of the time, and when, for instance, depression mimicked Alzheimer's in older patients, both might be labelled "dementia."

A perfect example of this confusion arose from the dementia caused by syphilis. Over the course of fifteen to twenty years after infection, the disease spreads to the brain and results in loss of tissue in the frontal and temporal lobes and a dementia-like syndrome called "general paresis," "paralysis of the insane" or "paralytic dementia." Syphilis first appeared in significant numbers after the Napoleonic Wars, and I've seen estimates that it was responsible for at least half of all cases of so-called insanity throughout the nineteenth and early twentieth centuries. However, syphilis-induced dementia wasn't actually distinguished from other dementias until 1874.

Yet for this disease, unlike others, a treatment was eventually developed. In 1927 the Nobel Prize for medicine was awarded, for the first and only time, to a psychiatrist, Julius Wagner-Jauregg, for his invention of a radical approach to this late-stage syphilitic complication. Having seen instances of fevers interrupting or even eliminating psychoses years before, in 1917 he inoculated a thirty-seven-year-old patient with blood from a malaria-ridden soldier. As the malarial parasites multiplied, they caused periodic fevers; after six such attacks, the patient's neurosyphilitic symptoms faded, and he was eventually discharged from hospital, apparently completely well. The fevers had killed the syphilis bacteria. Wagner-Jauregg reported nine such cases the following year: nine patients who, if they had gone untreated, would almost certainly have died demented. The Nobel followed nine years later.

Wagner-Jauregg's story is significant in that it demonstrated that for some diseases of the mind, treatment was possible. Also, some argue that because this condition was usually associated with damage to the blood vessels of the brain, that connection—blood vessels and dementia—held sway long after syphilitic dementia had been made irrelevant by antibiotics. This persistent belief helped establish the now mostly outmoded idea that dementia results from "hardening of the arteries."

So it's a complicated picture. The variety of dementias that have existed through the ages makes it difficult to discern which ones may have been Alzheimer's and which were induced by other means, such as syphilis. It's also uncertain whether large numbers of Alzheimer's cases simply weren't recorded or even noticed by the experts. And finally, because of generally shorter life spans before the mid-twentieth century, there were not as many people of advanced age who could have manifested the disease.

The American population in 1800 (European descent only) was a little over five million—fewer than the number of Alzheimer's patients in the United States today. I have not been able to find statistics estimating the number of people who were over sixty-five in 1800, but a hundred years later, when the U.S. population had reached seventy-six million, an estimated 4 per cent were beyond that age. Given the advances in health care that had occurred over the century, it's reasonable to suspect that less than 4 per cent of the population back in 1800 had reached sixty-five. If that number had been, say, 3 per cent, 150,000 Americans would have been sixty-five or older in 1800. If Alzheimer's had been as prevalent then as now, as many as 10 per cent of those, or 15,000, would have had the disease. Given that, at the same time, Pinel and Esquirol were just beginning to bring various kinds of dementia out of the shadows, it's not surprising that American medical experts were not penning pages of documents about a condition of advanced age afflicting only 15,000 persons.

By 1900, again because of rapidly increasing life expectancy, the 10 per cent incidence we see today would have created a much larger population with dementia: close to 300,000. (Maybe a little high but not much: I've seen an independent estimate of 160,000 to 260,000, together with the astonishing claim that in 1900 there were more cases of syphilitic dementia than Alzheimer's disease!)[7] But even then, if the prevalent attitude was that shrinking memory and growing confusion were simply a part of aging, it's not surprising that these larger numbers failed to arouse much interest.

What is remarkable at first glance is the apparent lack of medical interest in Alzheimer's disease through the twentieth century—after Alzheimer had made his crucial discoveries and while the numbers of Americans with dementia were growing into the millions. From 1910 to 1930, only fourteen articles on Alzheimer's were published in the two major American neurological journals, *Neurology* and *Annals of Neurology*.

What might account for this lack of attention? As we look back, the numbers of demented patients should surely have been growing at a consistent rate. Several things were going on: it wasn't clear that the cases described by Alzheimer could be applied much more widely, since his first case, the index case, was a relatively young fifty-one when she first came to see him, suffering from a condition that was typically found in much older people. And for years afterward, senile dementia continued to be set apart from

Alzheimer's disease. Alzheimer's was assumed to be a dementia that began in one's forties or fifties; "senile dementia" was something different, which affected people decades later. Autopsies were done only on people under sixty-five who had died suffering from dementia; no one with "senile dementia" was considered medically interesting enough.[8]

There was still the attitude that dementia—at least among the old—was simply a normal event, part of aging. Not inevitable, but never a surprise. And there were also competing explanations for the disease, ones that demanded investigators take a step back from the microscope and look more at the person than the person's brain.

By about the middle of the twentieth century, a significant number of psychiatrists were not persuaded that dementia was a brain disease marked by the signs Alois Alzheimer had identified: the microscopic junk in and around brain cells, the plaques and tangles. This view was justified to a certain extent by the fact that autopsies had revealed an inconsistency: sometimes the brain of a demented person was untouched by these abnormalities and sometimes a person who had been cognitively intact at the time of death was revealed to have had a brain full of plaques and tangles. This lack of a one-to-one correspondence led some to look for other factors. The most ardent proponents of one approach, called "psychodynamic theory," suggested that dementia was caused largely by social factors, like the isolation experienced by many people over sixty. Here is how one of the foremost proponents of this theory—David Rothschild—made the case in the abstract of a scientific paper:

> *The results of a pathologic study of 24 cases of senile psychoses are reported. . . . A search for physical factors that might*

modify the normal aging process and be responsible for the development of a senile psychosis has proved unproductive. There was a lack of correlation between the histologic changes and the degree of intellectual impairment. Attention was also called to the fact that equally severe alterations may be found in the brains of old persons of normal mentality. These inconsistencies were attributed to differences in the capacity of different persons to compensate for the cerebral damage. The view is expressed that this capacity rather than the pathologic changes themselves is the determining factor in the origin of a senile psychosis . . . and the suggestion is advanced that the inability to meet personal problems may be the crucial factor. . . .[9]

Although somewhat obscured by the technical jargon, Rothschild's message is pretty clear: social factors, not biological ones, lead to dementia. There were others who put the case more strongly by arguing that mandatory retirement, loosening of family ties and lack of social stimulation could actually create brain damage such as faltering circulation, which in turn would lead to dementia: "Lonesomeness, lack of responsibility, and a feeling of not being wanted all increase the restricted view of life which in turn leads to restricted blood flow. . . . Clinical experience demonstrates the relationship of these psychological and sociological factors to the cell death of the senile. . . ."[10]

This notion that social causes were of primary importance was not circulating only in the psychiatric community. David Stonecypher, an M.D. in Boston, wrote an article in the *New York Times Magazine* called "Old Age Need Not Be Old," in which he denied that dementia was the result of the "physical deterioration of the aging brain" and argued that it instead stemmed from "the immense degree of *fear* and *frustration* associated with growing old." Much of the article detailed how retirement, losing financial security and losing loved ones created the symptoms of dementia.[11]

It's not so much that such views were completely wrong—indeed, there is a growing sense today that aspects of our social life do indeed affect the risk of getting Alzheimer's disease. But these ideas certainly impeded the progress of biological studies of Alzheimer's and also contributed to the sense that somehow Alzheimer's disease was not relevant—too rare and characterized by irrelevant abnormalities in the brain.

Even those who sought links between organic events in the brain and the symptoms of dementia sidelined Alzheimer's work for decades, principally among them researchers who claimed that Alzheimer's disease, or dementia generally, was caused by hardening of the arteries in the brain or "arteriosclerosis" or "stiff pipes." Before Alzheimer's discovery, some thought that all cases of senile dementia were caused by hardening of the arteries in the brain, and their belief was not weakened by his and subsequent cases showing the deposition of plaques and tangles. The notion persisted despite the existence of persuasive evidence that the brains of those who had dementia were not always compromised by inadequate blood supply. Oddly, even though these dissenting opinions were strongly expressed as early as 1910, the idea of diminished blood supply as the cause of dementia hung on, both in the public's mind and in medicine. Sir William Osler's *Principles and Practice of Medicine*, first published in 1892 and generally acknowledged to be the foremost general medical text at that time and for years to come, included Alzheimer's under the section "Senile Arteriosclerosis" until 1947.

In fact, while the idea of hardening of brain arteries hung on, a related idea got a huge boost in the 1940s from Dr. Walter Alvarez, an American physician renowned as one of the best internal medicine specialists in the country and famed for writing about medicine for the general public. Alvarez had a knack for putting things simply and directly, sometimes to the bewilderment of his colleagues. (Editors occasionally inserted technical terms in his writing, presumably because it was too readable.) In 1946 he published an article in the first issue of the journal *Geriatrics*, in which he argued that small, mostly undetected clots that interrupted blood flow in the brain were responsible for many cases of dementia. He referred to the process as a "piecemeal destruction of the brain," which, because of its measured pace, often went unnoticed. He wasn't reinforcing the idea of atherosclerotic dementia, as in Osler's text, but arguing that this was a hitherto undiscovered phenomenon that, to his mind, was very common. Not hardening of the arteries in the brain but the accumulation over years of tiny strokes.

It's not that Alvarez was ignorant of Alzheimer's disease: in this paper, he argues that brain autopsies could reveal the accumulation of tiny, but deadly, clots and the dead brain tissue around them in the same way that Alzheimer's is distinguished by its own markers. But he leaves no doubt that he thinks the clot story is hugely significant. It is an unusual article by today's standards for scientific journals because the writing is almost folksy and definitely absorbing: "I remember a fine old doctor who one day brought me his friend and neighbor of long standing, . . . [an] apathetic, prematurely aged man of 55. Glancing at the registration blank, I instantly got my hunch from the fact that the patient was listed as a hotel owner and manager. Evidently the man before me could not have taken proper care of a flophouse, so I judged that something must have happened to his brain."[12]

But the piece is more than 14,500 words long! Obviously, the writing was good enough to persuade people to read all the way to the end because some Alzheimer's experts claim that this one article influenced thinking about dementia for forty years (others settle for thirty). It persuaded both doctors and the public that blood clots—mini-strokes—caused dementia despite the fact that there was plentiful contrary evidence.*

It wasn't until the 1960s and 1970s that the burgeoning efforts to research dementia finally focused on Alzheimer's. The case for a major role for blood clots and hardening of the arteries was seriously eroded by research in the late sixties. A set of three brain autopsy studies established beyond a doubt that arteriosclerotic damage could be only a minor contributor to dementia, whereas the signs of Alzheimer's disease, plaques and tangles, accounted for something close to 70 per cent of occurrences of the affliction. True, poor circulation contributed to some extent, but its role paled in comparison to that of plaques and tangles. No matter how the numbers were assembled and compared, they supported the idea that the majority of cases of dementia, at all ages, were actually the disease described by Alzheimer at the beginning of the twentieth century.

Eventually, as so often happens in science, one publication, one strongly worded statement, turned things around. Just as Alvarez's article had propped up the idea of blood clots and dementia for decades, so an editorial by Robert Katzman published in the April 1976 edition of the journal *Archives of Neurology* declared that Alzheimer's and senile dementia were the same disease and that they ranked as the fourth or fifth most common cause of death in the United States. (At the same time, however, the illness wasn't even listed among the 263 acknowledged causes of death by Vital Statistics of the United States.)

---

* What goes around comes around: there is today growing evidence that *microscopic* damage caused by clots does indeed increase the risk of dementia.

Katzman didn't mince his words: the cellular damage noted by Alzheimer was indeed found to correlate with the severity of the affliction; there was no way to distinguish senile dementia from Alzheimer's disease except by the age of the patient. Then Katzman made the startling claim that there could be anywhere between 880,000 and 1.2 million Alzheimer's patients in the United States.[13]

His editorial had a huge impact on medical thinking about Alzheimer's, but two public events that happened at roughly the same time catalyzed the new awareness of Alzheimer's. One was the sad dénouement of the career of Rita Hayworth. Hayworth had been a huge star through the 1940s and 1950s. She was beautiful—her pinup picture was second only to Betty Grable's in popularity among troops during World War II. She danced on screen with Fred Astaire and Gene Kelly and appeared in films with the most famous actors of the time. One of her five marriages was to Orson Welles, but the one that is the most important in this context was to Prince Aly Khan.

Hayworth was the first celebrity known to have Alzheimer's disease. At one point, she was being threatened with expulsion from the New York City co-op where she lived because the other residents thought she was always drunk. A writer for the New York *Daily News* contacted Alzheimer's researcher Katherine Bick because he had heard that rather than being drunk, Hayworth had Alzheimer's. Bick told him, "I know absolutely nothing about Hayworth's condition, but I can talk to you about Alzheimer's disease."[14] The subsequent newspaper article chastised the co-op board for not knowing more about the disease. It wasn't entirely the board's fault; some people close to Hayworth continued to encourage her to appear in films (even though she couldn't remember her lines) and took her to parties. Finally, in 1979 she was diagnosed with Alzheimer's.

The second event that sparked public interest in Alzheimer's was a 1980 letter to the syndicated advice column "Dear Abby."[15] It was from "Desperate in New York," who asked, "Have you ever heard of Alzheimer's disease? I feel so helpless—how do others cope with this affliction?" The author of the column, Pauline Phillips (who wrote under the pen name Abigail Van Buren), suggested that readers with similar questions should forward them to the newly established Alzheimer's Association. Both she and the organization were shocked when more than twenty thousand letters arrived. (I've read conflicting opinions about whether or not "Desperate in New York" was actually someone with the organization, trying to create interest in it.)

That combination—an advice column and a New York newspaper story about the suffering of a beloved actress—forced Alzheimer's disease into the public mind, where it has been ever since. To follow up: Hayworth died from complications of the disease in 1987 at the age of sixty-eight. And Hayworth's daughter from her marriage to Aly Khan, Princess Yasmin Aga Khan, became a force as a fundraiser for Alzheimer's research.

Pauline Phillips contracted Alzheimer's herself years later and died in 2013 at the age of ninety-four, but her singular contribution to Alzheimer's awareness lives on as the Mayo Clinic Abigail Van Buren Alzheimer's Disease Research Clinic. On the medical side, Robert Katzman's editorial set Alzheimer's research on the course it still runs today, a course where scientists deal every day with the fruits of Alois Alzheimer's research, now more than a century old.

CHAPTER FOUR

The Case of Jonathan Swift

The actress Rita Hayworth had a huge impact on attitudes toward Alzheimer's disease in the United States. She put a face on it, a famous face. As did Ronald Reagan. But there is also an intriguing example from two centuries earlier: Jonathan Swift, the author of *Gulliver's Travels*. He wrote about something that sounds like Alzheimer's, and he himself might have had the disease (both claims are debated). It's notable, though, that the disease later known as Alzheimer's quite possibly existed in the early seventeen hundreds.

Gulliver's Travels was a runaway bestseller, but Swift was greatly irritated when he viewed the first edition because his publisher had made several changes. In fact, in Swift's view, the publisher had abused the manuscript. He shouldn't have been surprised by the altered text. The satiric work targeted too many important people—from government officials to the Church to the Crown itself. That's why the author had insisted on anonymity (it was *Gulliver's Travels*, by Gulliver himself, although at the time most realized the author was Swift), and that's also why Swift had the manuscript delivered to the publisher, Benjamin Motte, at night and by horse-drawn cab. Yet Swift was angered by what he saw as the dilution of the blunt force of his message.*

Many of the characters Gulliver meets on his travels, such as the Lilliputians and their giant counterparts, the Brobdingnagians, are now well-known if not household words, but he also encounters one little-known group, which sheds light on both Swift and senile dementia.

Toward the end of the book, Gulliver travels to Luggnagg, where he meets the Struldbruggs. These unusual people, who make up a small percentage of the total population on Luggnagg, are immortal, and when Gulliver first hears about them, he is moved to write pages on how wonderful it would be to be a Struldbrugg: "I would first resolve by all Arts and Methods whatsoever to procure myself riches. . . . I would from my earliest Youth apply myself to the Study of Arts and Sciences. . . . I should be a living Treasury of Knowledge and Wisdom and certainly become the Oracle of the Nation."[1]

Gulliver's innocence is quickly exposed by those Luggnaggians who are not Struldbruggs: they tell him how miserable the immortals are. Not only do they acquire the infirmities of old age; they also curse the fact that they can never die. The king of Luggnagg even wishes he could send a couple of Struldbruggs back to England so Gulliver could prove to his countrymen that their abject fear of death was perhaps misinformed. Unfortunately, this turned out to be against the law.

But there was more to these immortals than the perhaps sur-

---

* Imagine, then, how he would have reacted to the 2010 movie version of *Gulliver's Travels* starring Jack Black as Gulliver! Swift wasn't even mentioned in the credits (anonymous to the end!).

prising fear of *not* dying. They were "opinionative, peevish, covetous, morose, vain, talkative,"[2] and eventually, "they have no remembrance of any thing but what they learned and observed in their Youth and middle Age and even that is very imperfect. . . . The least miserable among them, appear to be those who turn to dotage, and entirely lose their [memories/understandings]."[3]

Was the right word in that previous line "memories" or "understandings"? "Memories" *was* the word printed in the first edition of *Gulliver's Travels*. But in Swift's own copy of the first edition, now in the library in Armagh, Northern Ireland, he has underlined "memories" and written "understandings" in the right-hand margin in pencil. No matter. Many changes were incorporated into the 1735 edition of *Gulliver's*, partly to accommodate Swift's complaints at the beginning of this chapter, but this wasn't one of them.

The difference between the two is significant. Swift could have been describing the loss of memory, especially recent memory, that accelerates with age and can develop into dementia. Not just absentmindedness; serious memory loss. As Gulliver said, they have "no remembrance." However, Swift may have meant to write that the eldest Struldbruggs were instead losing their "understanding," a much more serious disruption of thinking that would likely indicate dementia. Of course, he was writing about Struldbruggs, not real people, but the words have the touch of a witness.

And there's more: "In talking they lose the common Appellation of Things, and the Names of Persons, even of those who are their nearest Friends and Relations. For the same Reason they can never amuse themselves with reading because their Memory will not serve to carry them from the Beginning of a Sentence to the End. . . ."[4] Swift might have been a particularly insightful reporter of these things because he himself might have had dementia; he certainly suffered from some of the symptoms, but the question is whether the evidence is strong enough.

"I have entirely lost my memory," Swift wrote in 1738 at the age of seventy-one, corroborating what friends had already observed for years, and again, in 1740: "I can hardly understand one word I write."[5]

Finally, in 1742 Swift's condition was confirmed as a result of an official inquiry by Dr. J.T. Banks:

> *[He] hath for these nine months past, been gradually failing in his memory and understanding, and [is] of such unsound mind and memory that he is incapable of transacting any business, or managing, conducting, or taking care of his estate or person. . . . His understanding was so much impaired, and his memory so much failed, that he was utterly incapable of conversation. Strangers were not permitted to approach him and his friends found it necessary to have guardians appointed to take more care of his person and estate.*[6]

Swift died two years later.

The retrospective diagnosis crowd has been very active over Swift: Was he depressed, did he have a brain infection, was he demented? Even if he was demented, was it Alzheimer's? Besides Swift's late writings and a couple of commentaries by his contemporaries, there really isn't much to go on, although that hasn't held back the experts. In particular, Paul Crichton, writing in the Lancet in 1993, argued that the much less common dementia known as Pick's disease was likely what afflicted Swift. Yes, Swift's memory was failing, and yes, he was having difficulty conversing, but those are features of Pick's disease too. Crichton added that the presence of early emotional changes, like melancholy and "blunt-ing" and the absence of a couple of common signs in Alzheimer's—loss of control of movements and spatial abilities—suggested Pick's.[7] Swift's years of complaints of apparent creeping inabilities are also consistent with Pick's, but the long duration of whatever-it-was looks more like Alzheimer's.

Sadly, unlike the case of Auguste D., where there was actually physical evidence to be found, no part of Swift's brain was pre-served. A death mask, a cast of Swift's face, was made and has pro-voked lively commentary. (Given the disparity among diagnoses, most of them must be wrong.) One investigator saw the evidence of paralyzed facial muscles on the right side dragging the left side of Swift's mouth down and inferred from the odd appearance of his left eye that some sort of infection had taken hold of that side of his brain. Subsequent work argued that each of those conclu-sions was misguided.

In 1952 the neurologist Sir Walter Russell Brain (yes!) described that death mask in depth and found a truly remarkable set of symp-toms in its superficial countenance. He began by pointing out an injury to the left hemisphere of the brain, which would have accounted for both the apparent facial distortion (that previous writers had dismissed) and Swift's late-onset difficulties with lan-guage—and concluding, among many other things, that Swift was "an obsessional personality emotionally arrested at an immature stage of development."[8]

But Brain's speculative foray had been outdone by phren-ologists who actually got access to Swift's skull in 1835, when his body was exhumed to make way for repairs to St. Patrick's Cathedral. Somehow, his bones were available for study for ten days (!) before being reburied. Perhaps the most notable comment (for sheer bizarreness) came from a phrenologist who, noting a depression on the left side of Swift's skull, claimed that "the bones must have undergone considerable change during the 10 or 12 last years of his life, while in a state of lunacy."[9] An interesting idea— that "lunacy" can change the structure of the skull.

Swift's case illuminates the difficulty: retrospective diagnosis of even someone like Swift, the object of much public attention, is clouded by lack of physical evidence and the inevitable misunderstandings of comments made centuries ago. But at least with Swift, there's sufficient detail to allow us to puzzle, argue and conjecture. There are the close observations of his friends and, most important, his own thoughts about what was happening to him.

But as vivid as those descriptions are, they have never proven exactly what mental disorder Swift had. Even if they had, he was a lone example of a personal story in centuries of anonymity. We have no way of finding out whether his case was in any way remarkable for his time. Yet his Struldbruggs suggest that all the changes we associate with Alzheimer's today were familiar to Swift.

CHAPTER FIVE

The Biology of Aging

T

wo or three centuries ago, you could do science in your own home . . . well, not everybody, of course, but the wonders of the natural world were open to curious, inventive, imaginative amateurs, who in many cases had the huge advantage of no establishment to tell them what to know or to believe. Anton van Leeuwenhoek, he of the cascading wig, was an exemplar, a Dutch guy who built his own microscopes and both described and illustrated an astonishing number of never-before-seen treasures of nature: single-celled animals, spermatozoa, blood flowing through capillaries and, most famously, bacteria. He is "The Father of Microbiology." Van Leeuwenhoek is not exactly famous, but if there were a scientists' Walk of Fame, he'd be there.

Abraham Trembley would not but should be. He, like Van Leeuwenhoek, provided reports of his experiments to the Royal Society in London, but more important, he has been credited with inventing developmental biology and/or giving birth to experimental biology; either would be a great epitaph. And he gained that tremendous respect with his book *Memoirs Concerning the*

Natural History of a Type of Freshwater Polyp with Arms Shaped Like Horns, published in 1744.[1]

The freshwater polyp of the title is the hydra, a pond animal about ten millimetres long with a simple little tube of a body, one end stuck to plants (or glass, once captured), tentacles waving at the other end. Like a squid with its back end anchored to the surface. Hydras are ambush hunters of the most brutally efficient kind, capturing water fleas and other tiny organisms in their tentacles and stinging them to death. They're doing something right—they are extremely common in freshwater ponds everywhere.

Maybe it's because they never die.

Yes, they are immortal, and if it weren't for Abraham Trembley, we might never have known that. Although he didn't discover hydras (Van Leeuwenhoek did that years before—who else?), he kicked the door wide open to hydra experimentation by performing years' worth of careful experiments using nothing much more than a pair of scissors and uncommon dexterity.* He discovered the animals accidentally in water from a ditch and assumed they were plants. The species he collected were green and, at first glance, seemed immobile. Certainly, they were anchored to a surface. But as Trembley kept watching, he realized that the thread-like tentacles seemed to move in intentional ways, beyond simply waving in the currents. He then noticed that in a jar that had been placed on a windowsill, the polyps were all clinging to the glass on the sunny side of the jar. Trembley rotated the jar to put them all in the shade and watched as over the next few days almost all ended up back on the sunny side.

Then his growing doubts that they were actually plants

---

* Note that he was working at exactly the same time as when Swift's *Gulliver's Travels* was first published.

prompted the first of many crucial experiments. He reasoned that cutting one in half would be fatal for an animal but survivable for a plant. He did just that, creating two separate pieces of a polyp, and then, to his astonishment, watched over days as the head elongated to become a complete entity and the tail grew a head and became a second. "Who would have imagined that it would grow back a head?" he thought.[2] Despite his prediction that survival would have defined the polyp as a plant, Trembley concluded that it was instead a highly unusual animal and set out on a historic set of experiments. He cut the hydra in half across and lengthwise; cut it into fifty pieces; turned it inside out; and devised all manner of manipulations which revealed that these animals (as indeed they are) can regenerate from minimal numbers of cells.[*]

Trembley's work established hydras as brilliant experimental animals for studying tissue regeneration. More recently—that is, a hundred years ago—researchers who studied hydra knew that you could keep them for years in an aquarium without any visible sign that they were aging in any way.

What are they doing that we can't do? They keep building all kinds of new tissues, like the inner and outer cuticle, various barbs and spears to spike their juicy prey, muscles to expand the body cavity to engulf said prey or to propel the animal in a cartwheel. For hydras to be able to perform feats of tissue generation and/or reproduction like that, for them to simply keep on creating new individuals as trees create twigs, they have to be able to make all kinds of new cells *all the time* and new tissues from those cells, again, all the time. Never stopping.

---

[*] Oddly enough, the only time Trembley used the name "hydra" was to describe varieties he'd created with several heads. He invoked the many-headed Greek monster slain by Hercules only then, but the name has since been applied to the entire genus.

Our cells can't do that. Or rather, the only cells we have that *can* are stem cells, those that refuse to commit and maintain an almost embryo-like state from which they can differentiate into the specialized cells, some two hundred of them, that make up our tissues and organs. Early on, when we were embryos, we were fundamentally little balls of stem cells—they had the entire task of creating an adult ahead of them—but later in life, stem cells are much rarer, needed only for tissue repair, and as time passes, much less active.

Hydras, on the other hand, maintain their stem cell activity, apparently forever. That's why scientists are interested in them: even if they don't clear the way for us to live forever, they might at least shed critical new light on the process of aging.

They are admittedly pretty simple animals, a cylinder with an inside and outside layer of cells, a matrix sandwiched between. But don't be misled by that: they also have four kinds of nematocysts in their tentacles, cunning little mini-harpoons, some equipped with toxins, some with grappling hooks. They also have both male and female genitalia—they are fully functioning, independent animals. And whether they are simple or not, the fact that they can replace the entire inner and outer sheath of the body in a week is a testament to the exuberance of their stem cells.

But those cells can be shut down. When researchers in Germany were able to interfere with the expression of a gene called "FoxO," the hydra's stem cell activity slowed dramatically, causing the animal's reproductive rate to slow in concert.

That was a nice add to previous studies which had shown that people in both Japan and Germany who live to a hundred are much more likely to have certain versions of the FoxO gene than are people in the general population.[3] The importance of the hydra studies is that now there's a bridge, however slim, that connects a longevity-associated gene in humans (which seems to exert multiple effects) with an animal that never, ever dies. An animal which can be experimented upon without obvious concerns for animal welfare. (FoxO isn't the only gene that seems to influence the rate of aging. There are several. One of the most recent and dramatic demonstrations of their effects was an experiment reported late in 2013, where the life span of nematode worms was multiplied by five by combining the effects of two different "aging" genes.[4] That is, of course, the equivalent of a human living five hundred years. The researchers were expecting a 130 per cent increase, but they got a lot more than that. Another nice gene to have would be Klotho.[5] With it, you live longer and smarter. But one copy only— two copies diminish those effects.)

And while for my money, the biology of aging is truly one of the most fascinating kinds of science and could therefore be an end in itself, it's inevitable that any finding that emerges from it will be immediately seized on as new hope for extending the human life span. This isn't just a twenty-first-century dream. It may actually be eternal, but it had an excellent flourish in the late nineteenth century, at a time when religion was relinquishing its hold on attitudes toward aging and science was beginning to assert itself.

Curious individuals like Charles Asbury Stephens came to the fore. Stephens was a hugely popular writer with his stories of farm and village life in Maine, written for children in the magazine *The Youth's Companion*. The stories had young heroes and heroines and described in detail farm activities like beekeeping, maple-sugar making and cutting ice. *The Youth's Companion* was at one time the bestselling periodical in the United States. Stephens, though famous by his writing, wasn't content. He had calculated he had no more than twenty years to live and pursued a medical degree, supposedly so that he could write medical stor-

ies for the *Companion*, but these studies were also a nice tie-in
with his already well-developed interest in life, aging and dying.
Regardless, as soon as he had his medical degree, he started
turning out books on aging, with titles like *Long Life* in 1896
and *Natural Salvation* in 1903. *Natural Salvation*'s subtitle was
*Immortal Life on the Earth from the Growth of Knowledge and
the Development of the Human Brain.*

Stephens came to believe that aging was caused by a thousand
cuts, that we were brought down by tiny, incremental deficiencies,
wounds or failures and even by the mental fatigue that resulted. If
such insults could be delayed or prevented or countered, we might
live many decades longer. But exactly how to do that? Stephens
lamented that lengthening life might be just beyond his generation's grasp: "We can but feel, therefore, that we live at humanity's
darkest hour—the hour before the dawn. We live too late to be
buoyed and comforted by the illusions of religion, too soon to
reach the goal and snatch our lives from the grasp of death."[6]

There's no doubt Stephens felt that, at least in this arena, religion had been superseded by science: he dismissed the "ludicrous
flimsiness" of the evidence for a soul and argued that "prolonged
life is coming as a result of the increase of scientific knowledge in
every field."[7] The goal was prolonged life, not immortality, which
Stephens thought was likely impossible. Even so, extending life by
decades was no trivial goal. He did his best to gather the evidence
required to be able to prevent aging by converting his home into
a giant lab with working space for fifty people, but the research
effort never really got off the ground. He was in the scientific spirit
of the time, quoting Ilya Metchnikoff with approval (more of him
in a moment), especially regarding Metchnikoff's claim that life
could be extended forty to sixty years. Stephens suspected that
cells, the "millions of tiny artisans" of the body, held the secret—

that if it were somehow possible to purify the blood and thereby cleanse the body's cells, that would be a first step toward life extension. Stephens argued that the route to this reboot for the body would come from the energy possessed by the nervous system, and somehow, together, these approaches would extend human life thirty years, fifty at the outside. It wouldn't be quite the same kind of human life, though: what Stephens saw as the all-too-common "beast life" would be replaced by a higher version, in which greed and lust would fade as lives became longer.*

The notion that somehow the apparent inevitability of aging might be circumvented was given a push by two scientists roughly contemporary with Stephens. Ilya Metchnikoff, whom I mentioned earlier, was one. A Nobel Prize winner and acclaimed as the father of immunology (for his discovery of phagocytosis, the consumption of pathogenic bacteria by white blood cells). Metchnikoff was convinced that intestinal bacteria were to blame for the gradual breakdown of the body, culminating in death. It was a straight theoretical line from the accumulation of toxin-producing bacteria in the gut to the overstimulation of his beloved phagocytes, which in turn attacked a wide variety of bodily tissues, causing the gradual breakdown of age. (It must have been a heady time for bacteriologists—Stephens too argued that intestinal bacteria were to blame for much of aging.)

By promoting the consumption of sour milk, yogourt and other lactic acid–rich foods to clear the pathogens, Metchnikoff anticipated the current popularity of probiotics. He also foresaw one of today's cautions about extending life: he made it clear that it wasn't just about extending old age; it was also about ensuring that

---

* Another odd parallel: Stephens and Alzheimer were putting pen to paper at the same time.

health and the capacity to work and be productive would accompany that extension. At the same time, he seemed credulous in his acceptance of claims of great ages, like 152 and 185, having been reached in the past. In a sense, he needed to believe in those claims to allow himself to envision pushing back the commonly accepted limits to a human life.

Another scientist, the French Alexis Carrel, while making much less impression on the general public despite his 1912 Nobel Prize, is still revered for his experiments with tissue-cultured cells, living cells taken primarily from chicks and maintained in the lab for months, as much as twenty times longer than had been achieved previously. Carrel's chick-heart tissue even continued to contract for weeks after being cultured. He was sure that the end of even such prolonged survival was brought on by the accumulation of wastes and depletion of nutrients and so invented techniques for serially transferring cells into fresh media. In doing so, he broke all the records, keeping cultured cells alive for decades: chick embryonic heart culture number 725 lived an astonishing 34 years! These results predictably caused a public sensation and encouraged vivid sci-fi rumours: that the fragments of chick embryo heart had by now grown into a complete, beating organ; that the heart had to be constantly trimmed to prevent it from growing out of control; and—my favourite— that it was kept on a marble slab with groups of scientists tending or admiring it night and day.[8] The strange thing, of course, was that the cultured chick-heart cells lived much longer than a chick would have, persuading Carrel that cells had the potential for immortality, a secret that, if unlocked, could make possible the immortality of the whole organism.

Not surprisingly, at the same time that the science was suggesting the potential for prolonging life, quacks and frauds *promising* the same abounded, with testicular extracts in particular making fistfuls of money for their promoters. Whether or not the obvious desire for extending life was, as Stephens had thought, an inevitable result of the loss of religious comfort in the face of death, the passion for a longer life span began to flourish a hundred years ago and continues today, albeit with a completely different scientific foundation.

As far as we can tell, human life expectancy is greater now than it ever has been (assuming that biblical claims of ages in the nine hundreds represent nothing more than misplaced decimal points and that the claims in which Metchnikoff placed his faith were similarly exaggerated). Life expectancy is rising by roughly one year every four (although slightly faster in developing countries) and has been going up in some parts of the world since the mid-eighteen hundreds. Twenty-five additional years over the last century—not quite two days per week. The numbers don't all agree because there is a plethora of studies, each of which comes up with slightly different figures. The bottom line is that the increase in life expectancy over the last hundred years has been huge.

We're already seeing the results of increases from decades ago: more and more individuals are reaching what seems to be an upper limit of roughly 110 to 120 years,* and there are some six thousand centenarians in Canada. The most recent worldwide count of super-centenarians (more than 110 years old) was sixty-one, sixty of whom were women (although it's hard to keep up: one recent list is seventy-three women, two men, but then the world's longest-living man just died.)[9] Medicine, better nutrition and exercise take most of the credit for that, and at least for now, those

---

* The record as I write this is Jeanne Calment of France, who died in 1997 at the well-attested age of 122.

numbers appear to represent something close to a maximum, the human upper limit.*

But can that bar be raised? The advances gained in the modern understanding of the molecular mechanisms of life have inspired thoughts of stretching the human life span to 130, 140 or even 150 years or more.

Ray Kurzweil, engineer, futurist and author of *The Singularity Is Near*, sees aging falling prey to his forecasted trio of revolutions in genetics, biotechnology and nanotechnology. He has credibility, having invented the first scanner, the first print-to-speech reading machine and the first music synthesizer. He's a futurist and claims that 86 per cent of his predictions have come true. And he looks into the future and sees three major steps not just toward extending life but also toward actually making us immortal. Or, as he puts it, every year that passes will see more than a year's increase in life expectancy. That, according to Kurzweil, should happen in about ten years. The first major step is fully embracing what we already know about diet and good health. He himself takes about 250 supplements a day and thinks he hasn't aged much in the last fifteen years, though he's closing in on seventy. The second bridge will use advanced biotechnology to shut down disease processes and reverse aging. The third bridge, one that even he acknowledges is a little more remote, happening in about 2045, will see nanobots roaming around in our bodies, making sure that everything is going well. Then we can live forever. At the moment, Kurzweil is at Google, the company that launched its anti-aging project, Calico, in late 2013.

---

* It's true that life expectancy two hundred years ago was much less than it is today, but those who survived infancy, infectious disease and assorted other killers could still manage to live to ninety or even one hundred.

Kurzweil isn't the only one. Francis Fukuyama, in *Our Post-human Future*, dedicates a chapter to the "prolongation of life." Gerontologist Aubrey D.N.J. de Grey, a computer scientist and nonstop advocate of "engineered negligible senescence," a man who believes that aging is not just a risk factor for disease, but the *cause*, wrote in 2004: "I consider it highly likely that within ten years from now, if the rather modest necessary funding is forth-coming, we will have the ability to take a mouse cohort with a three-year life expectancy, when it is already two years old, and treble its remaining life expectancy (that is, give it a total life expectancy of five years)."[10]

The deadline for that prediction is upon us, with no such mice living out their extended years in a lab anywhere.* Yet while there's no consensus to date on exactly what causes us to die, no single biological mechanism that goes wrong or shuts us down, and something like *three hundred* theories of why we age, the scientific view of the twenty-first century suggests that we might be able to take steps to forestall what has, until now, seemed inevitable.

Rather than dwelling on predictions, we would benefit from taking a closer look at the science. For instance, looking back at Alexis Carrel's work through a modern lens changes it substan-tially. His "immortal" cell idea got a rude shock in the early 1960s when Stanford researcher Leonard Hayflick showed that cells in tissue culture actually had a built-in limit, depending on the source. It's called "the Hayflick Limit." Hayflick established that embryonic human cells would divide about fifty times, then die. Conversely, cells taken from individuals in their eighties or nineties

---

* It's worth reading more about de Grey—a controversial man. Check out Sherwin Nuland's portrait: "Do You Want to Live Forever?" MIT Technology Review, February 1, 2005.

divided only a handful of times before expiring. If cells were frozen and stored, then thawed and put back into a culture medium, they just picked up where they had left off, counted to their limit and then died. It showed beyond doubt that cells had built-in limits to their lives. But Carrel's experiments had suggested otherwise. What gives—or gave? No one really knows, but some suspect that every time Carrel introduced new culture fluid to his system, new cells were introduced too, and because Carrel had no way of telling new from old, he just assumed the cells were immortal.

The Hayflick Limit introduced a degree of precision into what had been until then a scattering of theories about why we age. There was the "rate of living" idea, which capitalized on the fact that the wide range of life spans in animals correlated well with measures of metabolism, like heart rate. A mouse's heart beats much faster than an elephant's; the elephant lives much longer. But that, in turn, raised questions: What is it about the rate of living that brings on aging and death? In the idea's original conception, there was a notion that we start life with a certain amount of "vitality," which, sadly, gradually drains away.*

Adopting an evolutionary point of view makes sense of it all. The driver in evolution is reproduction—that is what defines "success." Whatever combination of genes acts in concert to generate more offspring will be the combination that lives on into the future. That's, of course, natural selection, but the fact that it works through reproduction means that there is no mechanism to preserve genes that enhance life after reproduction (except perhaps those that provide for wise grandparent caregivers in humans). In this view, aging is simply that part of life not pro-

---

* Vitality had its predecessor in the mid-nineteenth century, when health proponents saw it as the "capital" you were born with.

tected or maintained by genes—life when the hands are off the wheel. Slight insults to the system that might have been ameliorated or even prevented before can now accumulate. Repair systems that ensured survival are susceptible to damage themselves, and eventually the whole system begins to break down. Worse, the genes that enhanced reproduction in early life may possibly be the ones that, once reproduction is complete, turn out to be deleterious. Regardless, the bottom line is that reproduction is more important than longevity in order to preserve a particular species.

All of this is more relevant to us than to wild animals, which rarely live long enough to worry about how they are going to age. There are some curious exceptions to this rule: Trembley's hydras, of course, and also some closer to our size, like tortoises, lobsters and sturgeon, all of which seem not to age. The ocean quahog, *Arctica islandica*, takes a good stab at immortality: it lives for centuries. One dug up off the coast of Iceland was 374 years old. It's the world's longest-lived noncolonial animal (that is, an animal whose cells are differentiated into different types, rather than being a simple cluster of identical cells). Further down the aging scale are naked mole rats, which live twenty-five to thirty years, eight times longer than a comparably sized mouse. Other than the idea that hydras are just bundles of stem cells that can continue to replicate, there aren't yet good explanations for the longevity of all these creatures. And because most are not good lab animals, it'll likely be some time before those explanations will come.

So evolution provides the backdrop, but what, specifically, might create the Hayflick Limit? What causes cells to go through life with a built-in constraint on their ability to divide? Hayflick had no clue back in the 1960s, when he discovered the limit, but it's now clear that the major players are telomeres, little caps on the ends of our chromosomes that shrink a bit with every cell division. They apparently function to keep our chromosomes intact, something like the aglets on shoelaces, but with each round of DNA replication as cells divide and their chromosomes replicate, a segment of the telomeres is not copied and is therefore lost.

Although he doesn't get enough credit for it, the Russian scientist Alexey Olovnikov had a beautiful "aha" moment, something he called a "serendipitous underground brainstorm" that explained why this loss happens. Waiting for a train in a Moscow subway station, he somehow mentally transformed the train into DNA polymerase, the enzyme that crawls along DNA to copy it, and the track into the DNA molecule itself. Think of it this way: The enzyme has to travel *over* the DNA completely to copy it. If the train is the same, there's a problem at the end of the line—the train can't go any further, so the piece of track it's sitting on can't be copied. Nor could an equivalent piece of DNA. This scenario implies that each time DNA is copied, it loses that short piece.[11] A neat example of thinking about science using analogies.

Telomeres have no essential genetic information; they just protect against its loss. However, once chromosomes lose their telomeres completely, they become fragmented and stick together, cell division is disrupted and life, at least for those cells, ends. We do have an enzyme called "telomerase," which prevents the shrinking of our telomeres, but it's not active in all our cells, although it is commonly present in cancer cells, which are in some sense immortal.

You might think that all you'd need to do is activate telomerase and we would go on forever—and that idea is indeed being entertained. However, the fact that cancer cells also benefit hugely from having access to their own telomerase makes it clear that this enzyme would have to be applied with discretion. Circling all the way back to the hydra, it is likely, although not yet proven, that hydra cells contain active telomerase and so forestall the end of cell division.

Apparently, lobsters, one of those organisms that appear not to age, have high levels of telomerase in their tissue. The 2009 Nobel Prize for medicine was shared by Elizabeth Blackburn, Carol Greider and Jack Szostak for their work on telomeres and telomerase.

Does all this imply that telomeres are *the* answer to aging? Probably not, because we die before our cells have run out of cell divisions (when we die, we still have telomeres at the ends of our chromosomes). But the reduction of telomeres might contribute: repeated wounds in tissues where telomerase is not active could exhaust the local supply of wound-healing cells, and if that situation were multiplied throughout the body—in blood vessels, for instance—aging would be accelerated.

But if telomeres aren't the complete answer, what else plays a role? It has been known since the 1930s that the life spans of mice can be enhanced significantly by restricting calories in their diets. This is a robust result, proven again and again in the lab, but the impact of similar diets on humans is unclear, and similar experiments with rhesus monkeys have been confusing—to say the least. Two separate experiments have analyzed the results of twenty years of feeding rhesus monkeys a diet 30 per cent lower in calories than control monkeys. One produced no significant lengthening of lives.[12] The other did,[13] and now the two groups are working together to make sense of their data. That will be an important step because it's unlikely anyone will repeat a twenty-year experiment. At the same time, the middle ground of maintaining a reasonable weight might at least protect us from some of the diseases that accompany age.

A diet high in antioxidants has also been promoted as a way of delaying aging, as the idea of accumulated damage due to oxygen running wild in your tissues has always held a prominent place in theories of aging. However, the experimental results are inconclusive. The world's longest-living noncolonial animal, the ocean quahog I mentioned earlier, does seem to deal with damaging oxidative processes much more efficiently than its shorter-living relatives. But those naked mole rats that live to prodigious ages for a small rodent actually have lower levels of antioxidants and higher levels of oxidative damage in their tissues than mice. If mice are genetically engineered to remove the key molecules that reduce oxidative damage, most of the time their life span is not increased; similarly, boosting the level of antioxidants often fails to lengthen their lives. It's not just mice or mole rats either: various experiments with different animals are casting doubt on the simple assumption that reducing oxidative damage, especially by consuming large amounts of antioxidants, will lengthen life.

The ideas that I've sketched out by no means exhaust the list of theories about why we age. The process might result from multiple causes, and we might never figure out why some animals appear not to age. But to be able to dig deeper into the Alzheimer's puzzle, it's important to distinguish, as much as possible, between aging and the diseases that become more common with age. Leonard Hayflick has commented that "death by natural causes" is a rare diagnosis these days, and yet, he argues, many deaths are exactly that, the natural causes being the gradual breakdown of the body as a result of aging.[14] However, in contrast to the Hayflick Limit, which has remained consistent since he discovered it more than fifty years ago, the entire human being has been displaying something radically different. Not only are we living longer lives than we used to, but there seems to be no end in sight. We just keep living longer.

CHAPTER SIX

A Natural Life

first became interested in the biology of aging at a time when I shouldn't have. Registered in a Ph.D. program at McMaster University in Hamilton, Ontario, I did my best to avoid the topic I was assigned and spent my time exploring others, irrelevant to my degree but somehow more attractive. One of those was the biology of aging, and that's where I first became aware of the difference, as it was then expressed, between life *span* and life *expectancy*.

Life span was the natural limit to a life of any animal. A mouse might live four years, a grizzly or a bison twenty, a whale much more and that ocean quahog, *Arctica islandica*, off the charts. At the extreme other end of the scale, some mayflies, once out of their pupal case, live twenty-four hours at most. Of course, these are not hard numbers: there will always be those who die an untimely death and some who greatly exceed what seems to be the limit (except for those mayflies). But those outliers simply represent the natural scatter around the number that is typical for their species. In the same way, you might expect your new car to last about 250,000 kilometres, but some fall far short while others just keep on ticking.

That's life span, a kind of natural, built-in limit. Life expectancy, at least as I learned it, was a measure of how close to that
maximum most of the members of that species actually came. So
animals in captivity generally live longer lives than their compatri-
ots in the wild by avoiding the predation, food shortages and
diseases of natural life. For their wild cousins, life expectancy is
short, even though their life *span* is the same.

When it comes to humans, it has always been clear that even
in circumstances where most individuals died young, it was
always possible for some to live into their eighties or even more.
Especially in the distant past, life expectancy was short, sometimes
just enough to have reproduced, and the human life span remained
an impossible dream for most. But something remarkable has
been happening since the mid-nineteenth century, a spectacular
extension of the average human life that represents such an enormous—and unanticipated—increase in human life expectancy
that it challenges the notion of the fixed life span. To get a feel for
just how remarkable this change is, we should go back to the late
seventeenth century.

Edmond Halley was one of an extraordinary group of scientists in England at the time. Elected to the Royal Society (to which
the best among them belonged) at the age of twenty-two, Halley
was a mathematician and astronomer, of course best known for
the brilliant realization that the comet of 1682 was in all likelihood
the same comet as the one that had been seen in 1607 and 1531.
Armed with Newton's ideas about gravity and with the concept
of elliptical orbits, Halley made the bold prediction that this same
comet would reappear in 1758, when he would be 102. He didn't
quite make it, but the comet did and forever after bore his name.

But like his fellows in the Royal Society, Halley applied his
skills to a wide range of disciplines, and in 1693 he published "*An*

Estimate of the Degrees of the Mortality *of Mankind, drawn from* curious Tables *of the* Births *and* Funerals *at the* City *of* Breslaw *with an Attempt to ascertain the Price of* Annuities *upon* Lives."[1] Halley was trying to figure out a rational route to designing annuities, income payable to individuals until they die. The issue, of course, is the risk, either to the insurer or to the insured. How long is this individual likely to live? The insured wants to protect against the money running out; the insurer wants to protect against paying out more than was taken in.

Halley was dissatisfied with most of the life expectancy data available for London and Dublin because the numbers weren't sufficient, ages at death weren't properly recorded and the data were muddied by the fact that both cities experienced a large amount of migration. If, for instance, young people moved away after birth, the age-dependent statistics would be messed up. Halley argued that the ideal setting was one where most of the population stayed put, being born and dying in the same place. Then the statistics from Breslaw, the capital of Silesia (now Wroclaw in Poland) fell into Halley's hands, and they looked ideal. The city had very limited population migration, in or out, and so Halley set about to analyze the demographics of the city.

He created a table of births and deaths at every age for every one of the residents from 1687 to 1691. During that time, there were 6,193 births and 5,869 deaths. But the timing of the deaths tells the story of human life in the late sixteen hundreds, a tale that acknowledges the deadly presence of infant and childhood disease. Halley showed that on average, there were 1,238 births every year. But only 890 of those infants made it to their first birthday. Of the survivors, 198 died over the next five years, so a mere 56 per cent survived to the age of six. From the safety of that perch, things got better. Handfuls of deaths were recorded over the next decades, the numbers slowly rising from half a dozen in the teens to eight, then nine per year until the age of fifty. After that, even though the population had been much reduced, the annual number of deaths continued to climb to ten and eleven. Finally, with very few citizens left, the death rate gradually fell to zero.

Halley was able to use the life tables he created to calculate life expectancy at any age (the odds of a man aged forty living another seven years? 5.5 to 1), the likely number of men fit for military service at any time and the annuity information he had set out to discover in the first place.

In a postscript, Halley added a couple of notable observations.[2] He pointed out that it made little sense for people to complain that they would be wronged if they didn't attain old age and instead died an "untimely" death because fully half the population had died by the age of seventeen. Instead, he argued that his countrymen should "account it as a Blessing that we have survived, perhaps by many Years, that Period of Life, whereas the one half of the whole Race of Mankind does not arrive."

Further, he noted that the number of women of reproductive age who actually gave birth was quite small—something like one out of six every year, where Halley suggested that four of six wouldn't be unreasonable. He reasoned that because "the Strength and Glory of a King . . . [was] in the multitude of his Subjects," celibacy should be strongly discouraged.

Things have changed dramatically since Halley's time. Infant and childhood mortality have been hugely reduced (at least in the developed world), and that reduction has been the major contributor to an astounding increase in life expectancy. The statistic that stands out above all others is this: since 1840, life expectancy has risen roughly one year in every four. In the twentieth century alone, life expectancy rose by twenty-five years! My father was born in 1909. Life expectancy then was about fifty years. My son Max was born in 1992—by then, life expectancy was seventy-five.

Even with such tragic and widespread losses of life as occurred during the two world wars and the 1918 influenza pandemic, the upward trend was disturbed very little. A graph of rising life expectancy from 1840 until today is virtually a straight upward climb. The early days of that rise are easily explained: the elimination (or at least dramatic reduction) of infant and childhood mortality curbed the wholesale loss of life in the early years of the type recorded by Edmond Halley. That trend continued well into the twentieth century, especially with the advent of antibiotics in the 1940s and vaccines for polio and smallpox in the decades after. In Canada in the early 1920s, tuberculosis killed 85 out of every 100,000 people. Today: only 1.5. Diseases of early infancy killed 111 of every 100,000.[3] Today such deaths are negligible.

But these changes continue to confuse. A life expectancy of forty years centuries ago does *not* mean that people lived only to the age of forty, but rather, given the vast numbers of deaths under the age of five, your chances—at birth—of surviving to forty weren't very good. Once you cleared that hurdle, life was more forgiving, and forty was not a particularly remarkable age to attain. There were eighty-year-olds in ancient Rome after all.

But what has happened since, especially from the early 1960s on, is less well understood. With the risk of early death substantially reduced, the continued rise in life expectancy had to come at the other end, in late life. And it has. The sheer numbers—one year more in life expectancy for every four years that pass—are amazing enough, but here is a comparison that makes the increase in life expectancy even more astonishing.

A study published in 2012 compared the life expectancies of people living in the most advantaged (or "low mortality") countries, like Sweden and Japan, with both extant human hunter-gatherers and chimpanzees, wild and captive.[4] Their conclusion was stark: when it comes to life expectancy, hunter-gatherers are closer to chimpanzees than they are to their human counterparts in developed countries. Looking at this comparison from the point of view of evolutionary history, about six-and-a-half million years ago human and chimp lineages split from a common ancestor. Since that time, the two populations have diverged so that chimps in the wild today have a life expectancy at birth of less than fifteen years, while hunter-gatherers, like the Hadza of Tanzania, come in at roughly thirty-two years. That is a substantial difference, but it developed over millions of years. The recent acceleration of life expectancy—again, in the most favoured living circumstances—has occurred over *decades*. There are several comparisons, each of which is more astonishing than the last: a Swede living in 1900 had a profile more like a modern hunter-gatherer than a modern Swede; only four of the eight thousand generations of humans that have lived on the earth have enjoyed this extension of life; hunter-gatherers at the age of thirty have the same probability of death as a Japanese person at seventy-two (the latter comparison prompting the authors to say, perhaps predictably, "seventy-two is the new thirty").[5] And finally, the rise of human life expectancy has equalled or even outstripped many of the impressive and highly touted gains achieved by manipulating the genetics, environment or both of laboratory animals like fruit flies and mice.

It didn't matter how the dynamics of aging were perceived: human life was getting longer. Another lens through which to view the ultimate age is somewhat clumsily called "life endurancy."[6] It's an estimate of the fraction of a population that is likely to live to any age. Here are the figures if you take that idea to its extreme—that is, what maximum age will one person out of

a hundred thousand live to? From 1900 to 1980, life endurancy
for American men rose from 104.8 to 111.4, and for women, it
climbed from 105.4 to 113.

Such change made predictions irrelevant practically as soon
as they were tabled. In 1928 the statistician for The Metropolitan
Life Insurance Company, Louis Dublin, made the definitive pro-
jection of 64.75 years for a human life, underlining his confidence
with this addition: "in the light of present knowledge and without intervention of radical innovations or fantastic evolutionary
change in our physiological make-up, such as we have no reason
to assume." Although he didn't know it, women in New Zealand
had already surpassed his prediction.[7]

The rise in the length of an average human life has been so
meteoric as to lead to controversy. In 1980 Stanford University
physician James Fries looked at the current data and predicted that
the world would soon experience what he called "compression of
morbidity," morbidity being, depending on whose definition you
listen to, disease or anything that contributes to ill health.[8] That
is, as more and more people lived longer and approached what he
considered to be the human life span (roughly eighty-five years
of age), the length of time people would spend enduring illness
would diminish. They'd live full speed until crashing into the end
of life and dying, thus relieving the fear prevalent at the time that
society would be saddled with more and more enfeebled elderly.
In his view, chronic disease was being successfully postponed,
but life span was immutable, so what was formerly a slow decline
to death made unpleasant by illness became a full life with such
a rapid collision with the life span limit that the graph became
"rectangularized"—a little like Wile E. Coyote chasing the Road
Runner off a cliff, stopping in mid-air with his legs still flailing,
then falling like a stone.

How could Fries stick to his estimate of eighty-five as the human life span when many were (and are) living to one hundred? He saw them as the extended tail of the normal distribution, the natural scatter around that number I mentioned earlier. Inevitably, some fall short of eighty-five and some exceed it, but not enough in Fries' mind to persuade him that the human life span could be more than eighty-five. So sure was he that a subtitle in his article proclaimed, in bold: **The Length of Life Is Fixed.**[9]

At the time of Fries' article, life expectancy in the United States was seventy for men and seventy-seven for women. By taking the preceding gains into account and comparing the mortality curves, Fries settled on eighty-five as the maximum. He argued that life expectancy at birth was rising faster than life expectancy from age sixty-five and that they would meet at about eighty-five years of age somewhere around 2018. And that is why Fries was so confident that progress in curbing deaths by acute illness (such as heart attacks) and delay of chronic illness (which requires ongoing medical care) would create the compression of morbidity that he was predicting.

That was 1980. Fries and his group revisited the compression of morbidity thesis in 2011 with another two decades' worth of data in hand.[10] By that time, in the face of the rising tide of life expectancy, doubts had been raised about his unwavering support for a maximum life span falling somewhere in the mid-eighties, and Fries addressed some of those concerns. He argued that there were confusions between the oldest ages ever obtained, between 110 and 122, and the maximum possible *average* life expectancy, which he pegged at around 90. As he noted, "little is learned about a population from its outliers; it is interesting to know the height of the tallest man, but it conveys little information about human stature." He also claimed that too much attention was focused on life expectancy from birth, rather than what was going on after the age of 65. Life expectancy increases from 65 on were nothing like the dramatic increases from birth that had been seen through the twentieth century.

Still Fries and his group had moved from their original life span of 85 to something hovering beyond 90, an age that is possibly "greater than 90 years and is almost certainly less than 100," and in fact remarked that their original focus was on lessening morbidity as the life span was approached, wherever that life span settled. No more "The Length of Life Is Fixed" in bold.

Indeed, evidence had begun to accumulate, indicating that if there is a hard and fast upper limit to the human life span, it's not really clear where it lies. First, evidence from Japan, one of the world leaders in life expectancy, suggested that while it was true that Fries' compression was happening—that is, that the spread of ages at death was narrowing, so more people were reaching old age but dying within a few years of each other—something else unexpected was going on. The whole graph was shifting to the right. So yes, there was still a peak with outliers on both sides, but that peak was now close to ninety years, not eighty-five. More than that: the shape of the curve was the same as in earlier versions. So compression wasn't happening anymore—the curve wasn't getting squeezed. The whole pattern of mortality in Japan was shifting upward.[11] The question is, Where will it end? We simply don't know.

It wasn't only the Japanese evidence that persuaded researchers that the concept of a fixed life span was now on shaky ground, but it was crucial. Ironic that an aging population, which seems a threat to a country's secure future, is at the same time a treasure trove for researchers of aging. And what seems to be emerging is a more open-minded view that there might not be a fixed human life span.

Some authors defend the unusual proposition that from 1900 to 1950, there was a fixed life span of about eighty-eight years, but then "something" happened (their words) and life span began to increase. This, of course, turns the definition of life span on its head, from an entity that is biologically limited to one that is changeable. And that mysterious "something" couldn't be biological or genetic—a century simply isn't enough time for that. Instead, the important factors are environmental (like vaccination and improvements in sanitation and nutrition). Simple enough to say, but the claim that these factors have changed the "intrinsic rate of bodily decay" is extraordinary. No less extraordinary is the accompanying claim that humans "seem to be able to modify life span." [12]

So that is where we stand today. There are some incredible predictions on the table. One is that 50 per cent of the children born since the year 2000 will live to one hundred years of age. The same researchers argue that to the traditional groupings of childhood, adulthood and old age, a fourth must now be added: very old age. They also claim that fears that this new cohort will depend completely on society for their existence are misplaced: a sample of the (admittedly small) group of Americans over 110 years old revealed that 40 per cent were independent. [13]

It would be hard to imagine that the upward trend in life expectancy could continue at the rate it has, in the same way that it would be hard to imagine someone running the hundred metres in nine seconds flat. There must be a limit somewhere—but where?

Some worry about the rising tide of obesity and diabetes, both of which could take the place of former, now treatable causes of death. Exactly how much we should be concerned about that escalation is somewhat unclear. What is clear, however, is that obesity is increasing even more dramatically than life expectancy.

It began in the 1980s, and in the United States today, two-thirds of adults are either obese or overweight. The extremely obese represent the fastest-growing fraction of that group, and this rise has omitted no one: all racial and ethnic groups, all socio-economic groups and all geographic areas of the country. And while the two-thirds figure refers to adults, the rise in child-hood obesity has been just as startling. The National Health and Nutrition Examination Survey (NHANES) estimated in 2000 that Americans consume five hundred calories more each day than they did in the late 1970s.[14] And the obesity "epidemic" is not limited to the United States.

That two-thirds figure includes both the overweight and the obese. Stripping away the overweight, recent numbers reveal that about 35 per cent of the American population is obese (24 per cent in Canada), and those numbers, while not increasing as rapidly as they did in the 1980s and 1990s, are still holding fairly steady.[15] The most recent Canadian data show that by 2019, 21 per cent of Canadian adults will be obese, and if the obese and overweight are combined, they will outnumber normal-weight adults in five provinces: Newfoundland and Labrador, Nova Scotia, New Brunswick, Saskatchewan and Manitoba.[16]

There is no doubt that obesity reduces life expectancy, largely because of the associated diabetes and cardiovascular disease. Estimates suggest that obesity could reduce life expectancy by anywhere from one-third to three-quarters of a year. But given that life expectancy has been rising by three months every year, this finding suggests nothing much more than a hiccup in the trend. Even so, such an impact is bigger than that of all accidental deaths combined. And further, as the cohort of obese children moves into adulthood, the health complications they suffer could affect life expectancy much more; one estimate suggests a reduction of two to five years.[17] But there's an uncertain area in current research: one recent survey has suggested that people who are overweight (not obese) actually live longer, a mystifying result that has not yet been adequately explained—or accepted.[18]

So here at the beginning of the twenty-first century, we really can't say much with confidence about how life expectancy will change in the coming decades. And the analyses and projections about life span and life expectancy that we've just looked at have generally not investigated one important question—that is, Would we all succumb to Alzheimer's if we lived long enough? To get a better handle on that, we have to select one particularly important, aging organ: the brain.

CHAPTER SEVEN

The Aging Brain

n a poignant article in *Discover* magazine in 2012, Robert Epstein, a long-time associate of the controversial psychologist B.F. Skinner, described how Skinner had aged.[1] Although Skinner was fifty years older than Epstein when they met in the 1970s, he still had a powerful mind. But apparently not as powerful as it had been. Listening to recordings of a debate in which Skinner had participated more than a decade earlier, Epstein felt that the former, younger Skinner was wittier, quicker, sharper than the still enormously smart man he knew.

So Epstein and a colleague, Gina Kirkish, analyzed a set of recordings of Skinner: the 1962 debate, a 1977 debate and a speech he gave in 1990, when he was eighty-six. Skinner died shortly after that last recording. His speech rate declined steadily over the twenty-eight years under study—from 148 words per minute, to 137, to 106. Even allowing for the fact that a speech might encourage slower talking than a debate, the decline was notable.

Epstein was making a straightforward point. Even if we experience the healthiest possible aging, our mental abilities, or at least some of them, will inevitably decline. This is something completely separate from the destruction brought on by brain diseases like dementia, and researchers admit it's often difficult to disentangle the mental decline of healthy aging from the early stages of Alzheimer's or any of the other dementias. That's one reason it's important to know more about the aging process and to learn more about what "healthy" aging actually is. Once you have some kind of baseline, it's much easier to recognize when things are going wrong.

Brain aging can occur on different scales and in a number of ways, beginning with behaviour. How do people of different ages compare on tests of memory, calculation, attention and decision making? As we age, we keep pace in some things, but we fall behind in others. What do you see in the brain images of young and old while they're being tested? That their brains deal with challenges in different ways; that patterns of brain activity change with time.

According to Epstein, when B.F. Skinner was in his seventies, he was forgetful (failing to recall conversations from days before) and unable to cope with technical detail, even from his own previous publications. His memory decline is something that most would recognize and acknowledge. But there might have been more going on: difficulties with technical subjects could mean the aged brain was having trouble juggling concepts and symbols or following an extended argument.

Skinner likely remembered much of his earlier life. That phenomenon is probably the most common effect of aging: recalling with clarity what happened to us when we were young while forgetting the events of three days before, standing in a room wondering why we just went there with a purpose seconds ago or forgetting where we left the car in the mall parking lot.

"Working" memory is the system that allows you to memorize a phone number, then walk to the phone and punch in the right numbers. It's like short-term memory but more complicated. And while there are competing models of how it might work, one of the most popular envisages a three-part system, a "phonological loop," which stores acoustic information, including speech; the "visuospatial sketchpad," which briefly preserves what has just been seen, and the "central executive," which oversees both and coordinates them. The phonological loop is further divided into a part that actually stores the sounds and a rehearsal system that keeps the memory alive. Without rehearsal, the memory would fade and disappear in a couple of seconds. Hence the repeating of a phone number to yourself after you hear it.

There's a limit to how much you can stuff into your working memory and still retain. Psychologist George Miller earned life-long fame for arguing in the 1950s that we can keep approximately seven different pieces, or "chunks," of information in our working memory at any time.[2] The definition of "chunks" is pretty flexible: 2014 could be four individual chunks, or just one, packaged as a date. It's easy to experience an overload of your working memory. This is what is happening, for instance, when you have to turn off the car radio to concentrate while you're trying to find a parking spot. Keeping things in mind is what working memory is all about.

Robert Epstein's testimony suggests that B.F. Skinner's difficulty with more technical topics meant he was experiencing the gradual slowing of working memory. You only have to imagine a conversation—any conversation—to comprehend how central working memory is to life. Words, sentences and paragraphs tumble out at great speed and must be held onto long enough for a person to compose and utter a reply. At the same time, the visuospatial sketchpad is keeping track of the surroundings, who last spoke and to whom the next comment should be directed. All the while, the central executive is shifting focus and matching incoming demands with the available neuro-resources. And that's just the process for a conversation.

Such pandemonium suggests that somehow connections have to be made to longer stores of memory. Otherwise, conversation simply wouldn't be possible. Indeed, while most of the content of working memory simply fades away, some of it is transferred into long-term memory, where it can reside for decades. But long-term—even lifelong—residence in the memory stores doesn't imply the preservation of an unaltered, true-to-the-smallest-detail representation of the original memory trace. The "crystal-clear" memories of summer at the lake when you were ten may be vivid but are unlikely to be completely accurate. By now it's a well-established fact that memories can be implanted, rehearsed, refurbished and otherwise modified from the initial recording. The more time passes, the less memories resemble the original event, although the salient parts can be quite astonishing in their detail.

Long-term and short-term (or working) memory are only two of several categories. For instance, "long-term" can be divided into episodic (or autobiographical) and semantic (or factual) memories. Episodic: I ran into Pierre Trudeau on the street once in Montreal. Semantic: I know that Sir John A. Macdonald was prime minister—but I wasn't there.

These two kinds of long-term memory are apparently separate; a famous case study by Endel Tulving at the Rotman Research Institute in Toronto made that clear.[3] He worked with a patient known as K.C., who had suffered a serious brain injury in a motorcycle accident, wiping out his episodic memory but preserving the semantic. K.C. could tell you how to get to his cottage but couldn't remember ever having been there himself. He could play the organ but couldn't remember how he'd learned. On the other hand, he remembered factual memories that he'd learned before his accident, such as the difference between stalactites and stalagmites. The most striking thing about K.C. was that, as far as memories for facts about himself and the rest of the world were concerned, and in terms of skills like playing the organ or pool, he was just about the same as others of his age. But he couldn't remember himself being in any of those settings. He could remember none of his personal past, nor could he imagine the future. The only thing his mind still contained was the present.

K.C. had suffered extensive damage to the brain structure called the "hippocampus," known to be crucial for laying down new memories. In this respect, he was like another famous patient, Henry Molaison, also known until he died by only his initials, H.M.[4] The brain "accident" H.M. had was the result of an experimental surgery designed to halt his debilitating epileptic seizures. The removal of his hippocampus accomplished that goal but left H.M. unable to form any new memories from the time of his operation in 1953 to his death in 2008. In apparent contrast to K.C., H.M. did remember participating in things like bike riding in Hartford, Connecticut, when he was growing up. The fact that both could remember factual information but just one could place himself in the past shows how dissectible memory is.

These two individuals, despite the terrible misfortune of their circumstances, are rich and rewarding cases for researchers, far beyond what I've sketched out here, but they make a simple point: there are different kinds of memories, apparently maintained by different parts of the brain. For instance, destroying or removing the hippocampus does not affect memories that have already been established. They formed in the hippocampus and then at some

point moved on to more distant parts. Such damage does, however, leave a person in an H.M.-like state, with no new memories.

Now contrast that extreme state of memory disruption with
the slow decline of memory in healthy aging. Most of us, as we
age, acknowledge that our memories aren't what they once were.
But the loss is not dramatic: for every day you forget your first
date, there are days when you can remember who played bass for
Blind Faith.* Fading memories are universal: part of normal conversation, ruefully accepted, but still not well understood.

So two of the types of memory I've already mentioned, episodic and working, become less able as time passes. Semantic memory, the memory for facts, declines very little, although sometimes
the retrieval of such a memory takes longer.

Recently, Cheryl Grady and a team at Toronto's Rotman
Research Institute designed a study to look at three kinds of memory in the same individuals at the same time, something that had
never been done before.⁵ In this study, the researchers dissected
memory even more finely by differentiating "episodic" from
"autobiographical" memory in this way: autobiographical memories are of times when you were right there taking it all in and can
put yourself back in that situation by recalling it; episodic memories encompass that notion but also apply to memories of experiences that are minutes to hours old (often in a lab setting) and
are unlikely to last in one's autobiography. So the day you parked
your car and it was swallowed by a sinkhole is an autobiographical
memory; where you parked the car ten minutes ago is an episodic
memory, with little or no chance of becoming autobiographical.

Grady's study provided a clear picture of what happens in the
brain as we age. Subjects were shown pictures with brief descrip-

---

* Rick Grech, but that was easy, I know . . .

tions directing attention to what was salient: "grandparents," "airplane" and "poverty" were examples. Different kinds of questions then followed, each testing different memory systems. So a question for semantic memory might be "Who flew the first airplane?" but the episodic version of that might be "What colour was the airplane you saw?" and the autobiographical version, "When was the first time you were on an airplane?" Choices were presented for the first two, and participants could simply push the right buttons to give their answers. There is, of course, no right answer for autobiographical memory, and in that case, subjects were asked to remember an event and estimate its vividness. All subjects spent enough time in a magnetic resonance imaging (MRI) machine for their patterns of brain activity to be recorded.

As expected, areas of the brain involved in recalling semantic (general knowledge) questions were roughly the same in young people (mostly in their twenties) and older people (in their sixties and seventies). But there were significant differences when it came to autobiographical and episodic memory. The older group's brain activity was not as spatially defined. It was spread out, recruiting remembrances that were, by contrast, idle in younger brains. This blurring, or spreading out, of brain activity, at least in these two kinds of memory, had been seen before, but the fact that all three memory types were being examined in the same individual in this study heightened the significance. Yes, in a single brain, some kinds of memory require broader activation of the brain than do others. And the activation is broader in older people than in younger ones, making it clear that this change occurs with age.

What really does happen in older brains? This process of extending the activated area of the brain suggests a kind of compensation, a recruiting of brain resources not needed in a younger brain. In some test circumstances, this can work: older individuals sometimes perform just as well as younger by rearranging the nerve circuits they once employed into new, more far-reaching connections. But this process might also be a kind of desperate, and ultimately futile, attempt to keep up. For instance, in experiments similar to this in which older individuals *narrated* their autobiographical lives (rather than having their brains imaged while they just *reflected* internally), elements of general knowledge played a more important role than specific personal pieces of information as compared to the case with younger people. The personal detail just wasn't there, having been replaced, in part, by general, semantic knowledge (which, as I've already mentioned, is often well preserved in aging brains). As a result, older people's personal memories are not as richly detailed as they once were, but the individuals just aren't aware of that fact. Whether this is a curse or a blessing, I leave to you to decide.

Also, unfortunately, memory is not the only mental function that declines with age. Aging brains take longer to process information as well. We're best—at least speedwise—when we're young. Also, older experimental subjects have more difficulty ignoring a distractor while concentrating on a target. And, surprisingly, older people make more bad decisions, especially concerning risk. In experimental situations, both adolescents (known risk takers) and adults over sixty-five make choices like banking on a one-in-ten chance of winning twenty dollars, rather than taking a guaranteed ten dollars. No one doubts that this happens outside the lab as well.

What underlies these decrements in function in the aging brain? One is a gradual loss of volume, as our brains actually get smaller with time. But not all over: some areas are more prone to volume loss than others. So, for instance, the prefrontal cortex (the part of the brain that was the target of the notorious psychosurgical procedure frontal lobotomy) loses about 5 per cent of its volume every ten years, and that's every ten years after the age of twenty! This is a notable exception to the fact that we know much more about how the brain ages after sixty than we do about its aging during any of the previous decades.

One slightly puzzling finding is that the hippocampus (the part of the brain that was destroyed in K.C. and removed in H.M.) seems not to degenerate as quickly as the prefrontal cortex, and some parts of the hippocampus appear to remain virtually intact over time. (Ironically, those parts are among the hardest hit in Alzheimer's.)

An even more surprising observation in the hippocampus is that the dendrites, the arms of the neuron that receive incoming impulses, seem capable of growing anew, even up to and beyond the age of ninety. The puzzler is that the almost universal decline in memory, especially in the ability to lay down new memories, points to inadequacies in the hippocampus. The mystery is resolved by a closer look, which reveals that the hippocampus as a whole does steadily lose a small percentage of its volume decade after decade, despite the tenacity of some of its parts. This loss, while not critical early in life, becomes problematic after the age of sixty, and by seventy this part of the brain is diminishing by 1 per cent of its volume every year.

To get a better grasp of the loss of brain volume, we need to take a closer look at the chief elements of the brain, the neurons. Each neuron has a main cell body, an oval sac stuffed with the cell nucleus and its associated machinery, but what sets neurons apart from other cells in our bodies is their extensions. One, the axon, thin in comparison to the cell body, can reach out huge distances (on this scale) to connect with another neuron. (It's estimated that the same six degrees of separation that apply to humans also applies to the neurons in our brains: none is more than six steps

from any other.) They make no physical contact, but they do have
close encounters: a mere millionth of an inch (three one-hundred-
thousandths of a millimetre) separates the axon from the neuron it
has approached.

An electric nerve impulse bullets along the axon, and when it
reaches the end, electricity turns into chemistry. Packets of molecules called "neurotransmitters" migrate to the surface of the cell,
fuse with it and disgorge their transmitter molecules, thousands
at a time. These molecules drift randomly across that minuscule
space and plug into receptors on the other side. The place where
all this happens is called the "synapse." Provided enough of the
right kinds of neurotransmitters are released, the second neuron
will be stimulated to trigger its own impulse. Sounds very mech-
anistic, links in a chain . . . but in a space this small, randomness
creeps in at every step, resulting in variability and inconsistency at
the end. Yet somehow the system creates the thinking brain.

The axon releases the transmitters, and another extension,
called the "dendrite," receives them. Even though neurons have
but one axon, they may have hundreds of the much shorter dendrites, each of which can have scores of transmitter receptor locations. It's been estimated that each neuron in the human brain can
communicate with ten thousand others at once. So each single
neuron has a massive capacity for receiving, weighing and trans-
mitting signals. And each is only one of tens of billions.

Obviously, the brain is very crowded. In fact, beyond the
eighty-six billion neurons, there are at least as many cells called
"glia," whose function is somewhat mysterious. Once dismissed
as mere support cells for neurons, they have now been upgraded
to active participants in the workings of the brain. When you add
the fact that neurons at, say, the back of the brain connect not
only to their immediate neighbours but also to those much more

distant, you can begin to imagine just how staggering the wiring is. The international project BigBrain has recently produced a map of the human brain at unprecedented resolution, virtually down to the individual cell. The challenge now is to make use of the image, to develop techniques of handling the massive amounts of data it contains.

The crucial point about the loss of brain volume with time, in the hippocampus and elsewhere, is that it's the result not of dying neurons but of disappearing synapses. We may be conditioned to think that neuron death puts us at highest risk of memory loss. This is at least partly because of age-old warnings that every bout of heavy drinking destroys a terrifying number of neurons, usually cited as a hundred thousand. But this is a bogus claim in more than one sense. First, there's no evidence that such a loss actually happens. Second, even if it did, the absence of that number of neurons would barely be noticed. There are eighty-six billion neurons in the adult brain. In our cerebral cortex (the "thinking" brain), even in the absence of alcohol, we lose, to natural causes, one neuron every second! Eighty-five thousand a day. That's just "normal." A loss of eighty-five thousand per day over the months and years amounts to a loss of not quite 10 per cent of your total brain cells between the ages of twenty and ninety.

So cells aren't the issue; synapses are. The overall rate of synapse loss through life has been estimated to be significant enough that if a healthy person were to live to the age of 130, they would have lost about 40 per cent of their synapses and would then be in exactly the same position as an Alzheimer's patient: demented, even without plaques and tangles (and this is a thought worth considering for those bent on extending human life into that range).

Loss of synapses corresponds to loss of communication between neurons. The number of such connection points is almost incalculable to start with (eighty-six billion neurons, each of which might connect to ten thousand other neurons), but the loss is inexorable. And other issues come into play, to be sure: the myelin-coated "white matter" of the brain declines too, compromising the efficiency of communication, but it's the loss of synapses that contributes most to the decline of the aging brain.

Here is where some really exciting work is going on, investigating the question of how synapses actually fail. What really happens? Do they break down, commit suicide, get overstimulated and fall apart? What happens? It's easy to imagine that the process might be straightforward, with synapses collapsing, dendrites pulling away from axons and cells being destroyed. But it's actually much more bizarre and fascinating than that. First, it helps to abandon the simple picture of a neuron, that stripped-down version depicting nerve signals passing down a simple axon and dendrites accepting impulses from the axons of other neurons. The real situation is more sophisticated: for one thing, the axon can branch at the end, complicating its role dramatically; similarly, but much more impressively, dendrites are to be found everywhere. There could be hundreds or even thousands, together turning a cell into a dense, bushy tree.

But the focus can be brought in even closer, to an individual dendrite. Imagine looking at it as it stretches like a pipeline from right to left. On the surface there are what appear to be buds, called "spines," haphazardly distributed along the length. From mouse studies, it appears that huge numbers of spines are being generated every time there's a learning experience (such as running a new kind of maze), but over time, the majority of these spines are lost.[6] Only a tiny percentage survive until the end of the mouse's life. But every few days, a tiny percentage are added to a much larger number that were permanently established early in the mouse's life, and these new ones also have a permanent existence in the mouse's brain.

The spines come in three different forms: "stubby," "mushroom" and "thin." Stubby spines are mysterious. Mushroom spines are robust and long-lived, and each one has a large head and thin stem. Thin spines are very different. They pop into existence, may pop out again, may return, may not. Animal studies have made it possible to look at the changes in thin spines in the prefrontal cortex of the brain over time, linking their fate to the changes in working memory brought about by aging. At the front of the brain, in humans and in non-human primates, there's a specific area called "area 46," which is essential to working memory. Activity in this part of the brain ramps up when a human or animal performs what's called a "delayed-to-matching" test. Imagine a four-by-four array of squares, a quarter of a chessboard. The sixteen squares are randomly coloured—either red or black. You're shown the array, and then it disappears, only to reappear together with a different arrangement of coloured squares. Your job is to choose either the one that matches the first or the one that doesn't. While you're mulling this over, area 46 is hyperactive. (Experiments like this have been done with rhesus monkeys, and with them, area 46 is also hyperactive.)

That hyperactivity diminishes with time, coincident with lessening ability to perform the test accurately. And in animals, that gradual loss of accuracy is paralleled by a significant loss of thin spines—as many as a third of them—with aging. This observation suggests that thin spines play a key role in working memory and that their gradual loss over the years contributes directly to the weakening of working memory. It's also tempting to interpret the ability of thin spines to grow, develop, shrink and disappear over short time spans as being parallel to the demands placed on working memory, where the ability to track, make sense of and act on fleeting sensory inputs is paramount.

Anything that would cause the numbers of spines to be depleted over time would obviously impact our ability to store new memories and keep our working memory up to speed. But such changes would be considered part of normal aging with one caveat: as neurons lose spines or synapses of any kind, they might become more susceptible to the kinds of changes brought on by Alzheimer's or any of the other dementias. The first sign of pathology in Alzheimer's disease is the loss of synapses; the death of neurons comes later. So maintaining the health of synapses and delaying their loss might be one route to delaying, or even preventing, the onset of dementia.

CHAPTER EIGHT

Plaques and Tangles

In the spring of 1906, after painstaking staining and preparation, Alzheimer finally looked down the barrel of his microscope at the paper-thin slices of Auguste Deter's brain. And what he saw was devastation on a vast scale. First, the number of neurons (the brain cells) was dramatically reduced from normal. But he was also struck immediately by two more very unusual features. One of them was this: scattered among the remaining nerve cells were numerous dark spots, which he called "plaques." Although their structure varied slightly, they were essentially dark, round deposits. As odd as the plaques were, they were no stranger than the second odd feature he discovered: "tangles"—twisted filaments, aggregates of stuff *inside* the brain cells, looking in Alzheimer's drawings like locks of matted hair clumped together with bits and pieces of cell detritus. Some scientists have described them as "flame shaped." They were residues, sometimes the only evidence that told him a neuron had been there before.

The fact that he saw plaques isn't so remarkable—those had been seen before. At roughly the same time, Oskar Fischer, a scientist in Arnold Pick's rival lab in Prague, had described several cases of dementia where autopsies had revealed brains littered with plaques. So they weren't unique. But plaques together with tangles, underlying profound dementia, were a new discovery.

That was the beginning. Even today, more than a hundred years later, plaques and tangles are the distinguishing diagnostic features of Alzheimer's disease. Yes, there are variations in almost every direction: dementia without tangles (as in the instance of Alois Alzheimer's second notable case, Johann F.), similar dementias with tangles but no plaques, demented individuals whose brains have neither, and cognitively intact people with both plaques and tangles. A mixed bag admittedly (one that persuaded psychiatrists in the mid-twentieth century that there was much more to the disease than just these cellular anomalies), but the majority of Alzheimer's individuals have both plaques and tangles in their brains. These two abnormal deposits prompt pivotal questions about the disease: How did they get there? How do they cause disease? And maybe most critically, Which one is more important? Or even: Is either of these really important? At the moment, despite shovelfuls of money being poured into research on these very entities, there are no completely definitive answers for those questions.

So plaques and tangles exist, respectively, outside and inside neurons, the principal cells (but by no means the only and not even the most numerous) in the human brain. The billions of them form networks that allow us to sense the world around us, make decisions, have thoughts, feel emotions—the totality of human experience. As we've seen before, the basic neuron can be outlined pretty simply: the central cell body, the long axon reaching out to other neurons, and the hundreds of dendrites accepting messages from yet more neurons. But like all stripped-down descriptions, this one omits some items of importance.

For instance, much of what is of interest to us in the current context happens on the surfaces of neurons. Those surfaces are dynamic—undulating, shimmering, looking not unlike a densely planted garden of exotic vegetation. The plants are actually an assortment of molecules involved in the neuron's day-to-day existence. Some of these molecules sense the approach of messengers, facilitating entry of some into the cell while rejecting others; some are fooled into granting access to hostile entities, like pieces of foreign DNA or viruses. Yet other molecules package materials from the inside and ship them out or slice them into pieces to allow them to become functional. And as I've already mentioned, one of the most important things happening at the surface of the neuron, right at the synapse, is the release or reception of neurotransmitters.

The vast majority of the molecules on the surfaces of neurons are proteins, either wholly or in part. Protein molecules are the workhorses of life: no matter what biological process you look at, it's a virtual certainty that proteins are at the heart of it. They are building materials that are assembled with the assistance of other proteins. They can be rigid or flexible, huge or tiny, rare or common. All proteins are created as chains of subunits called "amino acids," but within fractions of a second, the completed chain folds into unique shapes: sheets, coils or other permutations of three-dimensional geometry. They are full of tightly packed twists and turns folded on themselves to create lumpy, globular shapes precisely determined by the attraction or repulsion between individual subunits on the chain. And while researchers believe there's probably one conformation that is the most stable of them all, many—probably most—exist in several versions.

Now let's get back to plaques, the hallmarks of Alzheimer's. Plaques are aggregations of protein molecules known as "amyloid" that are dumped outside neurons. The name is unfortunate because amyloid means "starch." Rudolph Virchow, the scientist who named the material, was persuaded by the way it stained that it was starch, and by the time amyloid was discovered to be a protein, it was too late to rename it. Amyloid proteins have a three-dimensional structure that encourages stickiness, so they're most often found as dense aggregates, usually associated with some disease process (as they most definitely are in this case). While amyloid is the key component, plaques are also composed of heterogeneous junk (collections of protein, bits and pieces of broken-up axons, dendrites from dead neurons, and sometimes, around them, other kinds of brain cells).

Although they must be deposited outside the neuron to begin to form plaques, the amyloid proteins nonetheless begin their lives inside it. One serpentine protein molecule, common at synapses, weaves its way through the neuron's outer membrane until part is inside and part outside, with a short section stuck in the middle (embedded right in the membrane). It's called "APP" (amyloid precursor protein), and a piece of it is indeed a building block of amyloid. It's an intriguing molecule because, while its role in the healthy functioning of neurons is unknown, it clearly has not just one but several tasks to perform. Assembled in the cell body (the main compartment of the neuron), APP is shipped "rush" along the axon toward the synapse. In a tribute to nature's economy, this super-long collection of seven hundred amino acids is so versatile that once it slips into place through the neuron's outer membrane, it can be snipped into pieces at predetermined spots, releasing shorter versions that perform additional tasks.

What does the snipping? Enzymes (which are also protein molecules), and they are specifically manufactured for the task of chopping up APP. This is as it should be, but sometimes one of the products of this trimming is a piece cut out of the middle of APP—

the piece sequestered inside the cell membrane—which is called "amyloid beta" or "A beta" or "Ab." In any other circumstance, this might well be the most interesting thing about APP because enzymes don't generally attack parts of molecules that are embedded inside the cell membrane, but much more significant is the fact that amyloid beta is the main component of Alzheimer's plaques. This excised protein fragment folds in a way that makes it stick to others like it. Then it is released outside the cell, and a runaway process ensues, with perhaps a million such pieces sticking together to form a visible clump in the cell—a plaque.

A caution: this is the barest of outlines of what happens in plaque formation, and it conceals the mysteries that abound. Plaques contain within them other bits and pieces from the cells around them; they are frustratingly absent when they should be present; they can exist in significant numbers without having any apparent effect on the brain; and they may not even be the entity that should be targeted in any Alzheimer's therapy. However, even with all this uncertainty, plaques remain one of the two significant signposts in the development of the disease.

Tangles are the other. Unlike plaques, they form *inside* neurons. Again, they are composed of a protein—in this case, a molecule called "tau," whose role is to help create and maintain the shape of neurons. Recall that neurons are more arms and legs than bodies: a neuron's role in the brain dictates the length of its axon and the number and length of its dendrites. In order to form connections to other neurons and to create and elaborate the spidery processes that make these links happen, the neuron must have an internal skeleton, a network of supporting structures that, while providing stability, is also flexible enough to be able to change direction and elongate. In the neuron, this role is played by microtubules.

But they cannot fulfill this role on their own, and that's where tau enters the picture. Tau stabilizes microtubules; with more tau bonded to them, microtubules become more rigid and resistant to change. With less tau, though, microtubules can fall apart and re-form, allowing them to adapt to the immediate needs of the cell. When sturdiness is all important, more tau is good. But when flexibility is needed, it's not. For instance, flexibility is required in the fetal brain when neurons are dividing, connecting and creating new synapses, and the microtubule network needs to keep pace. Reversible cellular mechanisms exist in each cell, which promote either attachment or detachment of tau to or from the microtubules. During these processes, phosphates are added to tau when more brain flexibility is needed and subtracted when less flexibility is required. The process is called "phosphorylation." The fetal brain (which needs to be flexible) has an abundance of phosphate. Oddly enough, so do the brains of hibernating animals. It's not clear why hibernators ranging from black bears to hamsters phosphorylate their tau like crazy when entering hibernation. You'd think, if anything, they'd require less brain flexibility, not more.

Regardless, when this mechanism is properly tuned, it is reversible: the fetal brain gradually abandons its hyperphosphorylation as the brain matures; the same thing happens when hibernating animals emerge from torpor. But in the human brain, in mid- to late life, phosphorylating tau can become a one-way, runaway process and is a key component of Alzheimer's disease. As phosphate is added, tau is released from the cellular skeleton and collects in mass amounts inside neurons, eventually aggregating, leading to the visible tangles that Alois Alzheimer saw. Some of them are literally all that's left of the original neuron and are sometimes called "tombstone" tangles.

It is remarkable that the two features that Alzheimer identified (amyloid plaques and tau-containing tangles) are still the hallmarks of the disease associated with his name—remarkable in that he was working in a context a hundred years ago when some of today's most familiar mental conditions, such as bipolar disorder (then called "manic depression") and schizophrenia were just beginning to be characterized. But while he saw clearly that plaques and tangles were present in the brains of his patients, Alzheimer had no clue as to their exact roles, and he didn't know what sequence of events caused them to appear. How could he? We still lack that knowledge today, though a major research enterprise has been launched in an effort to find out.

Following on the conclusive studies of the 1960s that plaques and tangles were indeed the agents most closely associated with Alzheimer's disease, a competition was launched: scientists who believed that the disease process started with amyloid countered those who assigned the key role to tau. At one point in this decades-long debate, people joked that the conflict had become a quasi-religious sparring match between tau believers and amyloid beta believers, *Tau*ists versus *BA*ptists.

But obviously, it's important to know how amyloid and tau relate to each other—if they even do. Is one dominant? Are they somehow dependent on each other? If that relationship were clear, therapeutic approaches could be much better focused.

Those who believed that amyloid was the key ingredient hit the ground running with the amyloid cascade hypothesis. This is exactly what it sounds like: a hypothesis that as amyloid collects around neurons, it reaches a threshold; once that threshold is crossed, significant brain damage and accompanying dementia set in. But *how* this happens wasn't (and still isn't) clear. Could it be that somehow amyloid triggers plaques, then tangles, and the two then carry out the destruction in concert? Or might tangles simply be refuse left over as a result of the damage done by the plaques before them?

One study that supported the amyloid hypothesis performed PET scans on patients suspected of having Alzheimer's (twice over several months), to establish the rate of tissue loss. At the same time, a chemical known to bind to plaques was circulating in their brains. The results were clear: the areas of the brain that shrank the most were also those with the heaviest load of plaques. The authors had no doubts: "This provides support for the central role of amyloid deposition in the pathogenesis of AD [Alzheimer's disease]."[1] However, you could argue that because the researchers were looking only for plaques and not for tangles, tangles might also have been found in the areas of greatest cell loss, a possibility casting doubt on the postulate that plaques were doing the most damage.

Indeed, the findings of this study ran counter to those of another, which showed that the degree of dementia exhibited by the patient was related to the number of tangles in the brain but not to the number of plaques. The authors also pointed out that the tangles had accumulated in the parts of the brain known to have been hit first and hardest, while the correlation between plaque deposition and those brain areas was not as strong.[2]

This is actually one of the most intriguing and challenging aspects of Alzheimer's disease. On the one hand, tangles accumulate where the greatest brain damage is seen, and the pace of their increase parallels the most striking behavioural deficits, like failing memory. On the other hand, plaques seem, at least at first glance, to be scattered randomly, even meaninglessly, around the brain. At least that's what has been thought until recently. But now evidence gathered over the last few years is showing that the places where plaques settle are known as the "default" areas of the brain.[3] These are large, well separated areas that appear to be the most active when our brains are idling—when we are not actively engaged in reading or puzzle solving or conversation. When we're daydreaming perhaps. Those are the default areas, and they are the primary territories for the deposition of plaques in Alzheimer's disease. How weird! The two diagnostic pathological deposits, which in some ways are likely joined at the hip, have their own peculiar, unique patterns of distribution in the brain.

I could quote numerous studies that fall on one side or the other of the plaques versus tangles debate: that's why there are competing theories and that's why there is uncertainty about the relative importance of these two signals of Alzheimer's. This is not merely an academic debate. If an effective treatment for the disease is to be developed, it has to be targeted toward the agent of the disease. And these days, it's beginning to look as if the ultimate target might be *neither* plaques nor tangles.

How could that be possible? The key idea is that the development of Alzheimer's is a process. There is pretty good evidence that we all start accumulating at least some plaques and tangles much earlier than you might expect, perhaps as early as age thirty or forty. It's not inevitable that they reach the levels diagnostic of Alzheimer's disease; it's not even certain that if you *do* have that many of them in your brain, you are going to be demented. Nonetheless, the fact is that both are the final products of a long chain of production and aggregation. Plaques arise from the dumping of amyloid from neurons; tangles arise from the accumulation of excess and nonfunctional tau inside them. What's not so clear is whether or not these end products actually trigger the damage of the disease.

Numerous pieces of evidence suggest that once you see either plaques or tangles in the brain, it's too late: the damage has been done or at least is well underway. It's not yet possible to paint a clear picture of what goes on, but if you pull all the current research together, the scenario might look something like this: brain cells begin to produce excess amyloid beta, which exits the cell and starts to aggregate in the extracellular space. This amyloid is not yet in the form of plaques; it exists as simple pairs, triplets or short strings of a dozen amyloid beta molecules. While there's still controversy about the importance of these plaque precursors, it seems as if they're capable of causing damage at nerve synapses on their own. (And synapse loss, not neuronal death, is the first thing that happens in the aging brain.) For people who are at risk for Alzheimer's, subtle cognitive deficits can be detected years before they actually show overt symptoms. It's a little puzzling that one plaque precursor, called "A*56" (a twelve-unit string), peaks as many as twenty years before there are detectable symptoms of mental decline and becomes rare by the time the disease is apparent. What happens to them? If they progress from short strings to plaques (after all, they are sticky and would tend to do so), plaques would simply be the end of a protracted chemical process and not relevant to the establishment of a disease that was initiated many months or years earlier. Counting plaques would still be diagnostic, but attacking them might not be very helpful. In fact, some have suggested that breaking up plaques might be counterproductive, as it would release the shorter-chain versions of amyloid beta that are the truly destructive ones.

It's important to point out that many researchers would disagree with this view, and a number of trial Alzheimer's treatments have been predicated on either preventing the overproduction of amyloid beta or, further downstream, actually busting plaques. None has so far been successful.

If it's possible that plaques are merely signs that the damage has been done, can the same be said of tau—that its visible manifestation as tangles inside neurons is after the fact? Maybe it's not so much the tangles (those twisted strands of tau released from microtubules) that begin the collapse and destruction of neurons and especially their synapses. Perhaps it's the release of tau in the first place that starts the damage. There is some evidence for this idea—that some pre-tangle form of tau (a smaller, less complicated assemblage) might at least trigger the beginnings of neuron death. For one thing, even though the appearance of tangles in the autopsied brains of Alzheimer's patients tracks the progressive loss of neurons, the number of dying neurons at any point is usually far greater than the number of tangles. This fact suggests—but it's just a suggestion—that a smaller, undetected version of tangles might be involved. Perhaps when first liberated from the cellular skeleton, tau is destructive even before it creates fully formed tangles.

A more fundamental question can be asked: Is it the detachment of the tau protein from the microtubules that causes the problems or is it the fact that tau becomes toxic as it accumulates? It's of course possible that both are true: the loss of tau from the cell's skeleton and its accumulation in an already crowded cell may both be bad.

That's one final twist in this plot: How are amyloid beta and tau connected? How are they, as they've been called, a "toxic *pas de deux*"?[4] As I pointed out earlier, a case can be made for either to be the predominant factor in the establishment of Alzheimer's disease, although more researchers today lean toward amyloid as the crucial first step. But is there a scenario that fits the two together in a persuasive way?

Questions like these bring us to the frontiers of research, where information is gathered largely from genetically engineered mice. They're altered so that they have a human form of

the gene that makes tau or the human form of the gene that makes
the amyloid precursor protein or both. While the observations
based on these experiments are invaluable, they must be treated
with the caution necessary when extrapolating from rodents to
humans. That said, these mice have provided some intriguing
things to think about.

Is amyloid driving the entire process? Could be: mice with
exaggerated amounts of amyloid, whether by injection of plaque-
laden brain material or by genetic modification, also produce
the malevolent forms of tau, as if one triggers the other. But the
reverse is not true: provoking the accumulation of hyperphos-
phorylated tau does not increase amounts of amyloid.⁵ On the
other hand, mice lacking tau genes suffer much less amyloid-
induced damage, a result that suggests tau is an active participant
in causing damage.* Putting the two together is tricky, however,
because one of the best-attested facts about Alzheimer's is that
in the dementing brain, tangles and plaques show up in different
places at different times.

The take-away message at this point seems to be that both
amyloid beta and tau play crucial roles in causing Alzheimer's disease, but it's still not certain whether that role should be assigned
to the final products (the plaques and tangles that Alzheimer
saw) or to some harder-to-identify upstream molecules. These
are all-important issues: given the cost of developing new drugs,
it's crucial to have selected the right targets for them. And it's
even possible that targets other than these two might turn up.
More about that in the discussion of anti-Alzheimer's drugs in

---

* It's worth asking how mice can survive when they're genetically engineered to make
  no tau at all. After all, tau supports the neuronal skeleton. Yes, young tau-free mice
  seem normal (at least to us), although older mice become aggressive and exhibit memory deficits. Whatever protects young mice apparently fades with time.

Chapter 14. But while it is true that plaques and tangles are the drug targets of choice, there is some disquieting evidence that the relationship between them and Alzheimer's is not straight-forward, that they may accumulate in the brain and *not* cause disease. Some of the clearest evidence of this theory comes from a unique study of nuns in the United States.

CHAPTER NINE

"I only retire at night"

I only retire at night." Those words were spoken by Sister Mary when she was 101 years old. She is one of the treasures of what's known as the Nun Study. In the 1990s Dr. David Snowdon and a research group teamed up with the School Sisters of Notre Dame, who maintain convents in several states in the U.S. In effect, the nuns agreed to be put under the microscope. They'd have their lives measured, their minds challenged, and in the end, their brains autopsied. They'd be part of a long-term study, so many of the oldest wouldn't live to see the fruits of their labour. But the knowledge gained could reveal a lot about aging and dementia.

Six hundred and seventy-eight of them signed up, Sister Mary being the first to agree to donate her brain after death. She had the qualities to give the study an immediate human face. She was a little contradictory: in some ways a late bloomer (she didn't get her high school diploma until she was 41), she nonetheless started teaching when she was 19 and did not retire from the classroom until she was 84. She was tiny (four foot six, eighty-five pounds) but not shy. Some even described her as "bossy."

Four months before she died, Sister Mary did her regular set of psychological tests.[1] Here's how she did: on the Boston Naming Test, which required her to name line drawings of things: 9/15. A related test requiring names of real objects: 8/12. Then she took a fluency test requiring her to come up with as many names of animals as she could in 60 seconds. She named 8, but had managed 7 within the first 15 seconds, all common animals, like cat and dog. Then she ran out of steam.

She recalled three or four out of ten words after having them read to her and scored 27 on the Mini–Mental State Examination (MMSE), when her expected score was 4. Yet she had less formal education than 85 per cent of the Sisters (and generally, the higher the education, the lower the risk of dementia). She actually did better on these tests than some who were ten to fifteen years younger (that is, ninety and eighty-five!). Her results are even more impressive when you consider there's something called the "terminal drop," an abrupt decline in health a few months (or even years) before death. So Sister Mary was likely labouring more than usual when she was tested. Yet her mind seemed clear to the end, and amazingly, on the whole, she showed little decline over the last ten years of her life.

So, a rare example of almost perfectly healthy aging: alert, happy, attentive, smart. But when Sister Mary's brain was examined, the researchers were shocked: it was full of plaques and tangles, especially in the hippocampus and part of the cerebral cortex. Classic signs of Alzheimer's disease: she should have been demented.

Her brain was small too—only 870 grams. Of course, she was tiny to begin with (seventy pounds at death), so that fact, together with whatever effects the plaques and tangles were having, might account for tissue loss and a tiny brain. Yet she never eroded cognitively.

This is why Sister Mary is the Poster Sister for this project.

If you were to look at her brain tissue under the microscope as Alzheimer looked at Auguste Deter's, you would be seeing much the same scene that he saw. But Sister Mary's brain, in contrast to poor Auguste D.'s, was intact. How had she staved off the damage? Was it because her brain was free of any significant damage from other causes—no pieces of dead brain tissue killed by a blood clot—or was it something else?*

Sister Mary was not the first example of the contrariness of Alzheimer's, where you expect the disease and it doesn't show or where you don't and it does. But she is one of the most vivid. The fact that her brain resisted almost certain illness is an eye opener. It's not ridiculous to ask the question, How did she do it and how can we bottle that? Of course, that's why the Nun Study and others like it are so important. Without them, we wouldn't get to know Sister Mary—and her brain—so well, and it would be that much harder to put the pieces together.

Of course, there were others. Sister Matthia lived to 104 and had plenty of tangles in her brain but was undemented. Sister Marcella seemed to be a replica of Sister Mary. She did even better on the MMSE (28 versus Sister Mary's 27), scored 8/10 on the delayed word recall and died at 101. But unlike Sister Mary's, her brain was clean.

Sister Bernadette had anything but a clean brain. She was in the ninetieth percentile for the number of tangles, which tracks the

---

* There is some direct evidence that people who have plaques and tangles in their brains but who are cognitively normal (like Sister Mary) show evidence of ramped-up activity in the neurons that would be most affected. They are bigger, they could be growing new extensions and they appear to be fighting back—either repairing the damage caused by the accumulating plaques and tangles or rerouting nerve circuitry to adapt to damage already done. See J.C. Troncoso et al., "The Nun Study," *Neurology* 73, no. 9 (2009): 665–73.

progression of Alzheimer's better than anything else. Yet she was unaffected as far as anyone could tell. She died of a heart attack at eighty-five. So we will never know whether she would have declined into Alzheimer's or continued as she was.

The cases I've presented here are celebratory, nice examples of people somehow overcoming the threat of dementia and living a good and healthy life, but of course, they represent only one side of the Nun Study. From the time it began in the 1990s, many nuns have died demented. It was in the course of looking for reasons why some succumb and some don't that the Nun Study team made a surprising, even shocking, discovery: the death of a Sister at age ninety from Alzheimer's was somehow connected to her writing skills when she was twenty. There's some kind of influence that arcs across time—seventy years of time. And it all begins with essays.

Beginning in 1930, young women of the School Sisters of Notre Dame who had been in the convent for a few years and were preparing to leave and begin teaching in the community were required to write a short autobiographical sketch. It had to be written on a single sheet of paper and was to include something about their birthplace, parents and education; interesting or crucial life events; and an explanation of what influenced them to enter the convent.

These essays have been evaluated for two things: idea density (how much information was economically packed into a sentence) and grammatical complexity (the number of embedded clauses in those sentences). The two tap into different brain mechanisms. Idea density is a subtle feature of expression that is measured as the number of ideas expressed every ten words.[2] Here's the sort of sentence you'd find in these essays, although this one's fiction:

I was born in Detroit Michigan on March 20th, 1913 and was baptized in St. Aidan's Church.

Seven ideas are nested together in this sentence, including "I was born," "born in Detroit, Michigan," "born on March 20th 1913," "baptized," "baptized in church," "baptized in St. Aidan's Church" and even "born ... and baptized." That is 7 ideas in 18 words, or 3.9 ideas in 10 words, amounting to an idea density score of 3.9.

By contrast, the following passage is much more idea dense—again, a fictional version:

The most joyful day of my life so far was my First Communion Day, April 11th, nineteen hundred and fifteen, when I was a mere eight years of age, and four years later in the same month I was confirmed by Bishop B. McRoberts. In nineteen hundred and twenty-one I was graduated from the eighth grade and it was only then that I could entertain the possibility that my desire of entering the convent was soon to be gratified.

You don't need a calculator to know that the second essay is very different from the first. And indeed, the idea density score here would be between 8 and 9, significantly higher than the previous 3.9. The Nun Study has found that the lower the idea density score in these essays written by twenty-year-olds, the higher the likelihood that those individuals would eventually get Alzheimer's disease. It sounds bizarre, but in the case of the two fictitious essays above, the author of the first would have been much more likely to die of Alzheimer's than the second. Overall, those nuns who went on to develop Alzheimer's scored an average of 3.86, and those who didn't had an average score of 4.78. It seems unbelievable that short exercises in self-expression can predict mental health seventy years later, but in the case of the Nun Study, they could and they did. As an example (although there weren't a lot of subjects), 90 per cent of those with Alzheimer's had recorded low idea density as compared to only 13 per cent of those who were disease-free.

Idea density is a curious thing: you write an essay when you're twenty-one or twenty-two and then put it away. Take it out many decades later, and it'll tally with your cognitive abilities at that older age. Another study has been done showing that idea density is an important factor throughout life: it tallies with brain pathology in older cohorts as well. But the really interesting questions would be: When does the ability to write with high density take shape? Have you reached your idea density peak by the time you're twenty-two? Or were you already at your peak when you were eight? Or, crazily, could idea density be taught? If so, it might be worth a few extra classes when you're growing up. The Nun Study did show that education was somewhat protective against the development of Alzheimer's, so maybe idea density is related to education.

The nuns' essays were also rated for grammatical complexity: the more clauses, the better. Grammatical complexity challenges working memory as you struggle either to follow someone else's edifice of a sentence or to keep your own under control. Each additional clause soaks up mental resources. You'd expect that as working memory sags with time, so does grammatical complexity. And that is true. But in this case, although grammatical complexity does decline over the years, it is not correlated with the chance of getting Alzheimer's. Only idea density has that mysterious relationship.

Alzheimer's aside, it's true that even if we're healthy, idea density and grammatical complexity both lose ground as we age. In one study, grammatical complexity and idea density appeared to decline slowly but steadily over sixty years. Here's an example of that phenomenon, based on two samples of writing by the same individual, the first from 1934, when the author was nineteen:[3]

> *In the year 1915 I opened my eyes to take a first wee peek at the world, just a tiny part of the immense world... I was baptized in _____ and was named _____ signifying pearl, certainly a foregoing of the pearl without price, my religious vocation. Two months later, leaving behind the city, bordering on the blue waters of Chesapeake Bay, we started for the glowing fields of . . . memories of childhood days, days of mirth and innocence. With joy I watched the new church and school rising gradually to the skies. And when I was six, the school was completed. What a coincidence! I entered upon my school career at the new school, _____. In May at the age of 8 I received my first Holy Communion. The following June found us on our way to _____.*

Then the same writer in 1995, at the age of eighty:

> *On _____, 1915 I was born in the memorable city of _____. A short time after birth I was baptized and given the name _____. My father was a carpenter and skilled in his work. My mother was a wonderful homemaker, taking care of us seven children and running the house very well. In _____ where I lived for eight years I received my First Communion in grade two. That summer, 1923, we moved to _____.*

The gradual but steady decline in both complexity and density was seen across the board, with participants who were otherwise well. Advanced education, even getting a Ph.D., didn't seem to moderate the decline in linguistic ability.

Of the two features, idea density is the more intriguing because of its connections to other phenomena related to aging. Besides the link to Alzheimer's in later life, other Nun Study research has shown that greater idea density at the age of twenty predicts a *longer* life as well as a healthier one. In addition, a question was prompted by the fact that these autobiographical essays deal largely with personal, emotional issues: Does the kind of emotion expressed (positive or negative) influence idea density?

Here's another pair of essays, one showing little emotion, the other much:

> *I was born on September 26, 1909, the eldest of seven children, five girls and two boys. . . . [M]y candidate year was spent in the Motherhouse, teaching Chemistry and Second year Latin at Notre Dame Institute. With God's grace, I intend to do my best for our Order, for the spread of religion and for my personal Sanctification.*

By contrast:

> *God started my life off well by bestowing upon me a grace of inestimable value. . . . The past year which I have spent as a candidate studying at Notre Dame College has been a very happy one. Now I look forward with eager joy to receiving the Holy Habit of Our Lady and to a life of union with Love Divine.*[4]

Judges rated the emotional content of essays like these, looking for either positive emotions, such as hope, gratefulness, accomplishment and love, or negative ones, such as sadness, fear, suffering or shame. Of course, in this case, given that the essays were written when the nuns were about to leave the convent and start teaching in the community, just prior to their final vows, they ought to have been happy! Maybe they were being careful to be upbeat, anticipating that the essays might be used to evaluate them. Nonetheless, those who expressed the most positive emotions lived roughly six years longer than those who didn't. It's worth noting that other studies have shown that cheerful, optimistic people often relinquish the longevity advantage they might have by taking more risks, like smoking and drinking. But, of course, this behaviour is much less likely to happen with the cloistered population in these studies.

There's no doubt that the most interesting and challenging observation provided by the Nun Study has been this link between idea density early in life and mental well-being decades later. But other interesting things have been discovered during the study of this unique population, including some that connect directly to physiology.

For instance, a number of Nun Study analyses have looked at the influence on brain health of infarcts. These are areas of dead brain tissue caused by blood vessel blockage and/or atherosclerosis (hardening and thickening of brain arteries), both of which lead to oxygen depletion in the brain. And given that the brain guzzles something like 20 per cent of the body's energy and oxygen supply, even a momentary depletion could have significant consequences. One study of the brains of 102 nuns, each of whom had a bachelor's degree and half of whom, a master's, showed that circulatory issues almost always made the situation worse. In other words, those who had plaques and tangles sufficient to qualify as an official Alzheimer's patient had always done worse on psychological testing if they also had infarcts—especially in areas of the brain thought to be critical for mental function. This result has been confirmed in other studies as well, with the conclusion that people with dementia often have issues with cerebral blood supply in addition to plaques and tangles. This is not to hearken back to the time in the mid-twentieth century when atherosclerosis was thought to be the *primary* cause, but rather to acknowledge that the damage underlying dementia is more often complex than simple. It's still not clear, given that complexity, whether preventing infarcts would prevent Alzheimer's. The plaques and tangles might have their way anyway.

Actually, there is pretty good evidence that they do. One way of gauging the spread of Alzheimer's is by using a scale of six stages, each representing a greater accumulation of tangles in the brain. This is a powerful scale because the accumulation of tangles seems to parallel the development of Alzheimer's fairly precisely. For some time, the first two stages were thought to be asymptomatic: though tangles accumulate in some parts of the brain, the individual who has the tangles exhibits no noticeable decline. But the Nun Study has made clear that this assertion is untrue: about 40 per cent of one cohort of nuns studied had measureable memory deficits, even though they were still in these first two stages. Findings like this have led to the belief that there is no chasm between normal mental functioning and Alzheimer's disease. Rather, normal functioning slides almost imperceptibly into mild cognitive impairment (MCI), which (not always but often) then moves into Alzheimer's.

There is a significant advantage to studying a group like this that is so homogeneous: the nuns don't smoke, don't marry, don't drink to excess, are the same gender, and are consistent with respect to occupation, income, living conditions, medical care and nutrition. They keep the same schedules and take part, at least when they reside in the convent, in similar activities, day in and day out. Many of the confounding variables that would render most studies like this unbearably complex aren't even on the table. Few groups that can be studied offer that kind of uniformity. The downside is that the nuns might have been a little *too* consistent, offering a view of aging that might not hold in a more diverse population. This possibility prompted a different group of researchers, at the Rush University Medical Center in Chicago, to undertake its own long-term study—in two parts. One part was the Religious Orders Study, essentially like the Nun Study but including men as well as women. The other was the Memory and Aging Project, which draws on a much wider population: men and women, Caucasians, Hispanics and African-Americans from retirement communities in eastern Illinois.

So far, some of the findings from the Rush studies back up the Nun Study, like the demonstration that many patients with Alzheimer's or its predecessor, mild cognitive impairment, have a mix of plaques, tangles, infarcts and other unusual protein deposits. It was discovered that even normal age-related decline tended to be associated with smaller numbers of these same abnormalities.

The Rush studies have also reinforced the observation that there is, at best, a blurred line between normal cognition, mild impairment and full-on dementia—a declining straight line—although the Rush studies have identified a sudden drop in cognitive ability about five or six years prior to the transition from one stage to the next. It makes you wonder how much of "healthy aging" is actually the stealthy creep of accumulated damage and disease.

That decline, even in apparently healthy individuals, occurs on an ongoing basis and accelerates gradually: an eighty-five-year-old slows verbally three times faster than a seventy-year-old, and motor, perceptual and verbal speed—all three of them—decline across all ages by an average of about 2 per cent per year.[5] But even with eight hundred participants over five years, these data are still all over the map, suggesting that individual differences are hugely significant and the causes might be an equally complex mix.

One factor does clearly accelerate cognitive decline, however: impending death, the phenomenon known as the "terminal drop." Postulates about this state of affairs are controversial, but they've been around for at least half a century, and they were supported in the Religious Orders Study, where there appears to be a sudden slump in cognitive ability three to four years prior to death. I mentioned this slump earlier when marvelling at Sister Mary's psych test results four months before she died. You have to wonder how much better she might have done if she'd been only in the middle of the terminal drop and not at the end.

The Nun Study established that education was an important factor in deterring dementia; the Religious Orders Study showed that education had no apparent direct effect on the accumulation of plaques and tangles but that it did lessen their *impact*. For each additional year of education, the effect on cognition was lessened —but only for plaques, not for tangles. (This result seemed clear despite the fact that an earlier Nun Study had linked low idea density specifically to the proliferation of tangles.) In fact, education's downward pressure on plaque impact applied to several different areas of cognition: it had an especially marked effect on perceptual speed and on semantic and working memory, but the impact on episodic memory and visuospatial abilities was lower.

However, the effects of teaching and thought on the accumulating damage to the brain are mysterious enough that the authors were obliged to say, "Formal education, *or something related to education* [italics mine], provides some type of cognitive or neural reserve."[6]

The value of taking the research out of the cloisters by extending the Nun Study into the Rush University Memory and Aging Project has become immediately obvious, as new factors affecting the development of dementia were unearthed. Among them were neuroticism (especially anxiety) and loneliness, both of which, especially the latter, might have been rare or undiscovered among the School Sisters of Notre Dame. However, people outside convents and monasteries who nonetheless maintained adequate social activity and support—or who had a strong sense of purpose in life—were also able to fight off the effects of plaques and tangles.

Here, the picture becomes both clearer and more complex. For instance, emotional neglect and/or parental intimidation and abuse experienced by children are associated with neuroticism: Does that imply that some cases of Alzheimer's disease were triggered by such issues when the individual was less than ten years old? Putting all this research together shows that something about the young brain (maybe even the very young brain) and the individual's experiences may result in unhealthy aging and possibly dementia. In most cases, the factors that have been identified are difficult to link to the eventual fate of the brain: for instance, the specific links between idea density and the health of neurons are not yet clear. How do a few years of education lessen the impact of plaque accumulation sixty years later? How can someone like Sister Mary, whose education was limited compared to that of the other Sisters, survive and thrive in the face of widespread brain pathology?

Somehow, the brain tissue must respond to these various influences in a way that slows the progress of the disease. To figure this out, we should know how Alzheimer's spreads in the brain. That's the subject of the next chapter.

CHAPTER TEN

A Deadly Progression

For my money, one of the most startling discoveries provided by the Nun Study was Sister Mary (and others like her), who managed to resist the destruction usually wrought by plaques and tangles and stay cognitively intact. The other was the examination of those essays written by women in their early twenties that could predict with reasonable accuracy how likely they were to get Alzheimer's disease. Set those findings in the context of the consistency with which plaques and tangles are found in Alzheimer's brains and important questions arise that beg to be answered. Here are two of them: Do plaques and tangles begin to appear early in life, earlier than most of us would have dreamed? And if so, what determines their influence on the mind if, in fact, most of us eventually get them? In this chapter and the next one, I'll look for answers to those questions, however incomplete they may be, given the complexity of the investigation.

Could those young essay writers with low idea density actually have had the very early beginnings of Alzheimer's disease? It's a natural question to ask, but one that seems not to be generally accepted by the experts, mostly because there's relatively little evidence that this might be the case. Instead, it's thought that somehow, by that early age, some type of vulnerability to the development of the disease has or has not been established (as represented perhaps crudely by low idea density).

However, it would be important to know when the pathology of Alzheimer's does begin. Can plaques and tangles or their precursors be found in the brain long before any signs of psychological decline are apparent? The answer to that question appears to be "yes."

The study that gives confidence to such an extraordinary claim examined the brains of forty-two people ranging in age from four to twenty-nine, male and female. Some had died in car accidents, some of cancer, some of heart disease, some even of strangulation. The list of causes of death was long—an interesting, young population in which to look for signs of Alzheimer's pathology. Can plaques or tangles be seen in those early years? While the answer is "yes," it is qualified. At least in this study, only one of the two telltale Alzheimer's signs *had* started to proliferate. When the researchers looked for signs of amyloid beta, or actual plaques, they found them in just one of the forty-two individuals under study. This person happened to have had Down syndrome, a condition that's prone to accelerated deposits of plaque. One out of forty-two, and that one at risk anyway.

The story was very different with tangles. They were extraordinarily prevalent, found in thirty-eight of the total forty-two brains! These bits of the protein tau, actually dubbed "pretangles," tended to concentrate in certain circumscribed areas of the brain. The stunning conclusion was that tangles may begin even before puberty: the four-year-old's brain was clear, but one six-year-old had some signs of tangles, and two eleven-year-olds were much more affected.

The authors of this study are Heiko Braak and Kelly Del Tredici.[1] Braak is a tangle man, celebrated, along with his late wife Eva, for their 1991 study, which established that only tangles, and not plaques, track the progress of Alzheimer's.[2] Plaques can be found almost anywhere, but tangles shadow the disease as it spreads through the brain.

This time around, Braak and Del Tredici push the very beginnings of Alzheimer's to an even earlier time, when there were no signs of tangles or pre-tangles at all in the place that is ultimately most heavily impacted in Alzheimer's—the cerebral cortex. In these brains, tangle material was usually limited to clusters of neurons tucked underneath the cortex. They had names like the *locus coeruleus*, the "blue place" because melanin granules give the cells in those clusters a bluish tinge.

Braak and Del Tredici argue that the results of this survey could only mean that tau, not amyloid, kicks off the disease process. Tangles, not plaques. This view would be just short of a paradigm shift if it were to be accepted. Despite boasting little or no success, antiplaque treatments based on the prevailing amyloid cascade idea have dominated attempts to develop medical therapies for Alzheimer's. Braak thinks going down that road is a mistake. It's obviously important to know where this process starts, and Braak and Del Tredici claim that it begins in those neuronal clusters.

But regardless of how justified their conclusion is (and it's a minority opinion at the moment), the observations alone say one really important thing: you don't have to be old to have what looks very much like early Alzheimer's pathology in your brain. Of course, this study of a young population couldn't establish that the early presence of tau in the form of pre-tangles would inevitably lead to Alzheimer's. But it provides data that must be taken into account when other findings are evaluated. And there are other findings that add weight to the importance of tau.

Nonetheless, the momentum behind amyloid beta and plaques has not stalled. Researchers at the Mayo Clinic, the University of Pennsylvania and a variety of University of California campuses broadened the picture by setting autopsy results to one side. Instead, they created a hypothetical timeline by bundling together several different ways of gauging the development of the disease in the living.[3] It's a bold attempt to depict the time sequence of Alzheimer's and the relative importance of a variety of possible contributors. They referenced a variety of so-called biomarkers: the rise and fall of amyloid beta in the cerebrospinal fluid, the rate of metabolism in the brain, amounts of tau and even the degree of brain atrophy revealed by MRI.

In perhaps their most important point, they assert that despite Braak and Del Tredici's demonstration of brains loaded with tau in the absence of amyloid beta, it is still possible to argue for a crucial role for amyloid, at least when it's en route to depositing plaques. They have on their side the observation, by more than one research team, that amyloid does rise and then fall in the spinal fluid of people who go on to develop Alzheimer's. This dynamic change in amyloid, the earliest signs of which could indicate the beginnings of the disease, takes place at least a decade before any symptoms appear.

But what does this group have to say about the already-established presence of tau, the other key diagnostic of the disease? The researchers argue for the independent development of the two by agreeing that tau is already present but not capable of causing Alzheimer's on its own. There is lots of tau present in mentally intact people, so perhaps it spreads slowly for twenty or even thirty years, but once amyloid beta takes off, it may be that tau immediately follows. Maybe the rise of amyloid triggers a new phase of tau deposits, or perhaps it removes some limits, such that the two proteins then head down the road to full-blown disease together.

Within this schedule, the researchers suggest that amyloid peaks before any disease is present, except perhaps the beginnings of MCI (mild cognitive impairment), acknowledged as a common prelude to Alzheimer's. Once that point has been reached, tau is still rising, and as shown by the Braaks, it is paralleling the destruction of synapses and then neurons themselves.

The limits of even this somewhat speculative study are that no population has yet been monitored through decades. Data have to be pieced together from a variety of studies, and so far, such data have tended not to focus on the two age groups crucial to filling in the puzzle of Alzheimer's development: middle age and late-stage dementia. The other shortcoming is that while the biomarkers in use are the best available, there is no guarantee that they include the most crucial elements. Brain changes might well be happening that are too subtle for today's tools, so we get only glimpses of what is really going on.

But one thing is clear: by the time the course of the disease has reached the point of overt destruction of the brain, it has taken up residence in the entorhinal cortex, a small area on each side of the brain, tucked under the massive cerebral cortex. The entorhinal cortex is a hub for neural messaging: it receives inputs from a variety of areas, including the hippocampus (the famed centre of memory recording), and it relays messages both back to the hippocampus and on to the higher centres of the cerebral cortex. The entorhinal cortex and the hippocampus have a close relationship: together they interact to allow us—roughly speaking—to remember where we've been and how we navigated to get there. It's not just about geography, as there's a deeper connection to memory of all kinds, but the important point is that the entorhinal cortex and the hippocampus are intimately related—in anatomy as well as in function. Disrupting the entorhinal cortex would clearly affect memory and then the mental functions dependent on memory, which, frankly, include most of them. One study established that in cognitively intact individuals, the entorhinal cortex houses something like seven million neurons, a number that is stable from the age of sixty to ninety. But as Alzheimer's takes hold, those numbers decline precipitously, in some places in the entorhinal cortex by as much as 90 per cent.[4]

Why is this part of the brain struck down first when more than eighty billion other neurons are available elsewhere? It could be that specific kinds of cells are present here that are particularly susceptible for reasons not yet understood. These cells may not necessarily even be neurons: one suspect is a kind of support cell called an "oligodendrocyte." It's responsible for myelinating neurons (that is, wrapping them in an insulating sheath of the fatty substance myelin). Myelination is an extremely important process, for that sleeve of fatty material must be placed around the axons of neurons to ensure that the transmission of impulses is swift and certain. It's curious that brain cells in the entorhinal cortex are the very last to become myelinated as the brain develops. Here, and in parts of the hippocampus, cells are myelinated long after those in the frontal cortex. Yet they're the first to fall prey to Alzheimer's. It is a mirror-image process. As Heiko and Eva Braak once put it, this "can hardly be attributed to chance."[5]

This odd inverse relationship has prompted a theory that Alzheimer's is really a disease of evolutionary history. It's well known that the process of wrapping insulation around neurons is complete in chimps long before the process is complete in humans, and it's perhaps notable that non-human primates never get Alzheimer's, no matter how long they live. Mind you, it's not really clear how this knowledge is going to help treat the disease.

There are also hints of a curious connection between the entorhinal cortex, where tangles first appear, and the default areas of the brain (mentioned in Chapter 8), where plaques are deposited early on in the disease. A study published in April 2014 showed that older adults with normally failing memories were unable to dial up the hippocampus while at the same time dampening the activity of the entorhinal cortex (something that appeared to be necessary for effective remembering) *if they had plaques* in the default network in their cerebral cortex.[6] This is one of the first indicators of some direct connection between plaques in the default network in the cerebral cortex and the beginnings of Alzheimer's relatively far away in the brain, in the entorhinal cortex.[*]

The catastrophic loss of neurons in the entorhinal cortex is unfortunately only the beginning of a cascading process in which diseased neurons in that location, before they become completely overwhelmed, join up with other neurons. The most important of these are in the hippocampus, that crucial instrument in the memory process where new and valuable information is first recorded, then passed on to places unknown to be stamped with a sort of permanence. But the disease does more than undermine the hippocampus, creating the faltering memory of Alzheimer's. It continues on, neuron to neuron, soon reaching the frontal lobes. And from there, it goes to the rest of the brain, leaving the motor and visual areas to the last. Once the process has been initiated, the spread of Alzheimer's is inexorable.

---

[*] A recent study showed that brains with plaque accumulating in them work extra hard in memory tasks, presumably to overcome the negative influence of the plaques. Elman, J.A. et al, "Neural Compensation in Older People with Brain Amyloid-ß Deposition," *Nature Neuroscience* 17 (2014): 1316–18. doi:10.1038/nn.3806

The prospect of slowing, interrupting or even stopping the disease seems daunting, but there might be a window of opportunity: How does the spread of the disease actually take place? Is it some kind of unfortunate coincidence whereby adjacent cells spontaneously begin to break down independently of each other—the result of a general environmental crisis in which they are embedded? Or do the crucial disease elements, the predecessors of plaques and tangles (or whatever they are), actually move from cell to cell, leaving a trail of destruction behind them? If the latter is the case (and the evidence supporting this theory is increasing), an opportunity does exist. Wherever you find traffic, there is the opportunity for roadblocks.

A growing body of opinion sees the spread of Alzheimer's disease as an analogue of the prion diseases: mad cow disease, the human dementia Creutzfeldt-Jakob disease and several others. As these are infectious diseases, they are of a different kind from Alzheimer's, yet they might progress in similar ways.

The infectious agent in the prion diseases is a malformed version of a protein native to brain cells. If introduced into a healthy brain, these malformed—actually *misfolded*—prion proteins are somehow able to cause their normal counterparts to misfold themselves. A runaway process of recruitment ensues, decimating the population of normals and swelling the numbers of misfolded until the brain can no longer resist. The misfolded proteins gain access to the brain in different ways. Some of the prion diseases are apparently triggered by the spontaneous misfolding of normal proteins. That is likely to be the case with sporadic Creutzfeldt-Jakob disease, which kills one person in a million per year worldwide. The mad cow epidemic in the 1980s, on the other hand, was anything but spontaneous: it was fuelled by feeding processed carcasses of dead cows (killed by mad cow) to living ones, inadvertently exposing them to the disease-causing prions. Only when such feed practices were banned was the epidemic brought under control.

As I say, these are infectious diseases, but the link to Alzheimer's might lie in the mechanism by which they spread. It's not yet clear exactly how misfolded prions turn normal proteins into virtual copies of themselves, but some sort of physical contact must occur. That would happen inside the neurons in the brain, but then a second step would be necessary: the misfolded proteins would have to be transported from one neuron to the next. And the next and the next. As some elegant experiments with both plaques and tangles have revealed, this is indeed the case.

Genetically engineered mice are the crucial players in these experiments. Mice are ideal lab animals (small, extremely fertile, easy to maintain), but they need to take some human genes on board to make them better models for human disease. For instance, when mice (engineered with human genes to develop plaques as they age) were inoculated with material taken from the brains of people who had died of Alzheimer's, plaques began to appear in large numbers and much earlier than they normally would have. Also, the plaques spread from their initial locations throughout the hemisphere of the brain that had been injected. It looked like transmission of the disease, but obviously this was a very special case because the experimental mice had been engineered to develop Alzheimer's later in life anyway. The injection may simply have accelerated—by maybe a few months—what was inevitable.[7]

But before long, this experiment was refined in a startling way: in this version, the mice carried human genes that did *not* predispose them to develop their own plaques. In other words, these mice would normally live a long and healthy life and maintain undiseased brains. Yet when inoculated with Alzheimer's material from a ninety-year-old woman who had died from the disease, the mice had fully developed Alzheimer's plaques in their brains after about a year and a half. It also seemed that early-stage plaques had started to appear much sooner than that, and plaques were found in parts of the brain distant from the inoculation site.[8]

While those experiments used full-grown plaques, scientists have demonstrated the same sort of multiplication and subsequent spread of just the main plaque component, purified amyloid beta. Stan Prusiner, the doyen of prion researchers, and his colleagues injected amyloid beta from the brains of diseased mice into the right hemispheres of the brains of healthy ones. They were then able to monitor the spread of the amyloid protein because the recipient mice had been engineered to generate luminescence wherever the protein aggregated. The amyloid beta spread from the right to the left hemisphere in the brains of the mice.[9]

The experiment was lauded by others as a beautiful one, but that wasn't enough for Prusiner, who couldn't resist a little propagandizing. In 1982 he had both coined the term "prion" and argued that it would hold the secret to many non-prion diseases, like Alzheimer's. He seemed determined to maintain or extend the hegemony of his term in a recent publication by referring to the Alzheimer's plaques as "prions," even though prions are infectious agents and plaques are not. Shortly afterward, the *New England Journal of Medicine* editorialized that the term "prion," with its obvious implication of infectiousness, should not be applied to Alzheimer's disease. The editors warned of the word's "dangerous baggage."[10] It remains to be seen whether Prusiner will respond to this.

So it seems well established that plaques (or the substances from which they're built) can spread in the brain. But how exactly? This is where the wizardry of molecular neuroscience comes into play. The more specific the question, the more abstract the experimental system must be. So to figure out how exactly plaque or tangle-like material makes its way from one neuron to another, it helps to take neurons out of the brain and grow them in cell culture. One experiment of this type showed that plaque precursors, short pieces of amyloid beta, could pass from one neuron to another, apparently directly, perhaps by being encased in a membranous globule and then leaving one neuron and being taken up by the nearest neighbour.[11] Shortly after that transfer happened, signs of damage began to appear in the receptive neuron. And while this experiment involved plaque-like material, the authors conjecture that this process might open the door to a similar cell-to-cell transfer of tau, the tangle protein.

If this were always true, it might explain the confounding fact that while tau tracks the spread of the disease in the brain, plaques do not. It may be "pre-plaques" that are doing the spreading. Such microscopic-scale events do need to be tracked so that we can get a fundamental understanding of what's going on. But will this help us develop a treatment? That is hard to say. It's extremely difficult to track such small molecules, at least in the whole brain, and by the time we catch up to them, they might already have done their damage anyway. However, microscopic observations have revealed molecular events that correspond to the behaviour of people afflicted with Alzheimer's. The early damage to the hippocampus, for instance, is reflected in the increasing inability to remember what just happened as opposed to events years before. And there are also other ways of correlating molecular events to behaviour. One surprising way involves the sense of smell.

While I was writing this chapter, a study was published showing that the ability to detect the smell of peanut butter declines perceptibly in the very early stages of Alzheimer's. This is not exactly news: many studies have shown that the sense of smell is less acute, even when Alzheimer's has just begun to manifest itself. But this recent study was striking because of its simplicity.[12]

Patients with either Alzheimer's disease, its precursor (mild cognitive impairment) or other dementias were asked to close their eyes, mouth and one nostril while fourteen grams of peanut butter, about a tablespoonful, was held in front of the open nostril. The patients were then asked if they could smell and/or identify the odour. If they couldn't, the peanut butter was moved closer in increments of one centimetre until they could. The results were pretty clear: on average, the Alzheimer's patients required the peanut butter to be ten centimetres (about four inches) closer than the other patients did.

Also significant in this study was the fact that the patients' left nostrils were much less sensitive to the odour than the right. This is consistent with the fact that the severity of the disease in its early stages is greater on the left than on the right. The centre for the perception of odour is among the areas known to be among the first sites to be affected by the disease. But we don't yet know what is responsible for this loss of the sense of smell. In mice, both amyloid beta *and* tau have been fingered as the culprits. There's no such detail known about humans, so much more needs to be learned, but this is a simple, low-tech way of determining the possibility of early Alzheimer's disease, something that will become more and more useful when there are effective ways of slowing the development of the disease.

Assemble all the pieces of evidence I've mentioned in this chapter and the picture is clear: slow, localized, cell-to-cell spread of pathological material is a good shorthand for the development over time of Alzheimer's. A time that should likely be measured in decades, not years. We may never know whether the Nun Study essays definitely represented an early link in this chain, but what they do point to, in addition to cases like Sister Mary, is that something established in early life can slow the development of the disease—as we shall see in the next chapter.

CHAPTER ELEVEN

The Brain Fights Back

Sister Mary is the most celebrated, but by no means the first, individual to die mentally intact having withstood a brain full of Alzheimer's pathology. In 1988 a team led by Robert Katzman revealed that plaques and tangles were fairly common in the brains of people who never suffered from Alzheimer's disease. (This was the same Robert Katzman whose strongly worded editorial in the journal *Archives of Neurology* twelve years earlier established Alzheimer's as a disease rather than a companion of normal aging.)

The team autopsied the brains of 137 residents of a nursing home (average age 85.5).[1] Of those, roughly half had been diagnosed with Alzheimer's, but it was the ones who weren't who were of most interest. Ten people who had performed in the top ranks in psychological tests turned out to have plaque counts that would have defined them as having mild Alzheimer's — up to 80 per cent of the counts of those with the full-blown disease. Eighty per cent and yet they were fine. It was studies like this one, and individuals like Sister Mary, that revealed cracks in the straightforward link between plaques and tangles and Alzheimer's. Thus was born the idea of "reserve": select people somehow had a little extra to give, some kind of capacity to resist those plaques and tangles.

The notion of reserve has gained enormous credibility in the last two decades, with report after report establishing that this is a real phenomenon, not a fluke of just a few patient populations. But even so, the meaning of the term isn't exactly clear. There's general agreement that there are two kinds: brain reserve and cognitive reserve. Brain reserve is more nakedly physical, found in a robust brain that is bigger and has more neurons and is therefore able to sustain damage without succumbing to it. Cognitive reserve refers to damage resistance afforded by a myriad of activities, like education, exercise and occupation. Of course, in the end these activities must somehow also exert their effects on the structure of the brain, maybe by creating larger neural networks or more diverse circuitry, but at this point, it's not clear exactly what those effects might be.

Intuitively, it would seem that brain reserve would provide better resistance. A bigger brain with more neurons should be able to function even as neurons are lost in the course of disease. (It's worth noting, however, that Sister Mary's brain weighed only 870 grams, and even factoring in her tiny body size—70 pounds at death—this is a small brain by any standard.) There is good evidence that brain size and neuron number do indeed exert protective effects, and the benefits can be demonstrated even if brain size is measured crudely.

A study published in the late 1990s measured the head circumference of nearly seven hundred people and found that those with diagnoses of probable Alzheimer's disease (seventy-five in all) tended to have smaller heads.[2] There actually was a trend with size: the 20 per cent with the smallest heads had more than twice the risk of the disease than the 80 per cent with larger heads, the risk being greater among women than men. But there was more: confining the analysis just to the people who had Alzheimer's showed that those with the larger heads still managed to do better on cognitive tests. The researchers weren't dogmatic about the meaning of these findings. They wondered if there might be some environmental effect that would both keep the skull small and predispose a person to Alzheimer's. Maybe, but they might have been witnessing a straightforward demonstration of brain reserve.

You might well question the validity of measuring the circumference of the head as a proxy for the size of the brain. (Or you might remember the controversial Philippe Rushton relying on tape measure estimates of head size among his students in order to back up his claims about the evolutionary relationships of the human races.) The brain and the head do not have the same circumference, of course, but the difference is a curious one. Our brains reach their maximum size when we're teenagers and then shrink, although very gradually, as we age. But our skulls don't shrink to keep pace because the bony plates of the skull have solidified. So measuring the circumference of the head at, say, age fifty is really doing personal archaeology: measuring the brain as what it once was rather than what it currently is.

Being able to do that opens up some cool experimental possibilities. Here's one. A group of Scottish researchers did a four-way comparison using intelligence-testing data gathered both in 1947, when a cohort of Scottish school children were eleven years old, and again when the same individuals were seventy-three years old.[3] Measures were taken of both current and former brain volume, using magnetic resonance imaging (MRI) to measure either the actual brain tissue (current) or the volume of the skull (former). Not surprisingly, childhood and adult intelligence correlated strongly. Current brain volume and adult intelligence were also linked—again no surprise—but remaining still was a small connection between maximum brain size in adolescence and current intelligence. This connection over decades caught the interest of the researchers. They pointed to it as evidence of brain reserve.

For those of us not obsessed with determining what is ultimately the most refined method for measuring the volume of the brain, some of the experimental efforts to do so might seem a bit surreal. In 1961, for instance, efforts were made to distinguish between the simple acts of measuring the head circumference, determining the volume of the skull and measuring the volume of the brain itself.[4] These anthropologists first took to immersing the back of the head of cadavers in water, up to the ears, and measuring the volume of water displaced and then repeated the process with the newly extracted brain. A three-way comparison of 105 corpses (mostly men, to avoid the complication of large volumes of hair) proved to their satisfaction that immersing the head is a better method because the errors that accrue are fewer than the errors brought about by measuring the skull's circumference. Of course, neither measurement takes into account the actual size of the brain itself. And the advent of magnetic resonance imaging rendered all these carefully designed techniques completely redundant unless one wanted to gather rough-and-ready data as quickly as possible without much expense.

The argument so far seems pretty straightforward: if you have more brain, you can afford to lose more brain cells. But as always, caution is recommended. For one thing, a bigger brain doesn't necessarily have more neurons. There are many other kinds of cells that could proliferate in their stead, although this argument doesn't figure prominently in the literature. In people with brain reserve, there could also be more *large* neurons. Indeed, Robert

Katzman's study (mentioned at the beginning of this chapter) noted a greater than normal density of unusually large neurons in the brains of cognitively normal people who had Alzheimer's pathology. And their brains were heavier than the average.

Taken together, these data support the idea of "brain reserve"—a hedge against dementia provided by larger brains with more neurons. It's not clear what combination of environmental and genetic influences might create larger brains, and sadly, by the time we become aware of their beneficial effects, our brains are already shrinking, so the usefulness of such knowledge, at least in terms of delaying or preventing Alzheimer's disease, is limited.

But "cognitive reserve" is different. It consists of a much more diverse set of factors, with the apparent therapeutic advantage that at least some of them might be influential throughout a person's life span. Education is the one that stands out: if there are studies showing that education fails to delay Alzheimer's, I haven't found them. And literally dozens of studies have established that the further you go with your schooling, the lower your risk of Alzheimer's. Those who stop at or before grade 8 have about double the risk of those who complete high school, who in turn are at greater risk than those who complete university and so on. Here, the population trends are encouraging: today, 83 per cent of Americans complete high school, up from only 13 per cent in 1910. And that upward trend was not confined to the early- to mid-twentieth century: in Canada between 1998 and 2008, the number of adults without a high school education dropped from 21 per cent to 13 per cent, while the number of adults aged twenty-five to thirty-four who had finished high school climbed to 92 per cent.[5]

In Shanghai, a 1990 survey revealed that women with Alzheimer's greatly outnumbered men. This finding isn't unusual, but the difference was stark in this case because many of the women had had *no* formal education.[6] And here's another example: in a recent review, the author, Yaakov Stern, refers to a study of 593 people over the age of sixty. Those with less than eight years of education had 2.2 times more risk of developing dementia than those who had gone further. But in this review, the author went on to look at much more than education, including the demands of one's occupation and a variety of leisure activities.[7]

Occupations were rated either as low "attainment" (including unskilled, semi-skilled, trades or clerical) or high (manager, business, professional or technical). Those grouped in the low-attainment occupations were 2.25 times more likely to get Alzheimer's. The same story was told through leisure activities: as long as the people in the survey participated in six or more, their risk of dementia was lower by 38 per cent. The activities ranged widely—from knitting or listening to music to working out, walking or going to movies or church. Not all of them seemed particularly mentally challenging; not all even seemed to be fitness inducing. But the merits of each weren't as significant as the degree of participation. Other studies of similar activities and education have arrived at the conclusion that by putting together every element that heightens cognitive reserve, one can reduce the risk of Alzheimer's by nearly 50 per cent.

As part of this review, Stern draws on yet more studies to make an additional point that might or might not be somewhat depressing, depending on your point of view. He cites evidence to support the idea that people with high cognitive reserve (who have a good education, an active mental and physical life and a challenging career) will nonetheless accumulate plaques and tangles over time, just as Sister Mary did. But unlike Sister Mary, most people like this will run out of reserve before they die and will begin to show the symptoms of Alzheimer's. When that happens, their decline will be much more precipitous than for those who lack such reserve. Why? Because by the time they show symptoms, they will already have accumulated a huge burden of pathology, and their decline will, perhaps ironically, bring them to the same point of dementia at roughly the same age as those who have been declining gradually for years in the absence of reserve.

Rapid decline is not something to be longed for, but a longer life without symptoms certainly is. So which would you prefer? Give me a long, symptom-free life and a dive off the deep end anytime, thank you.

So far, I've painted this subject in broad strokes of the brush, but inevitably, there are myriad details, and while it's not worthwhile mentioning every single study, there are some that surprise or put important emphases or qualifiers on what I've already said. Here's an unexpected one: bilingualism protects against Alzheimer's disease. Twenty per cent of Canadians and 50 per cent of the rest of the world are bilingual. Several studies have supported this conclusion, and the effect is substantial: bilingual patients admitted to one study for "probable" Alzheimer's were diagnosed more than four years later than their monolingual counterparts. In another, bilingual patients with Alzheimer's had more loss of brain tissue for a given cognitive loss.[8]

One of the beauties of bilingualism is that people are usually forced by their personal situation to learn another language, and often early in life, so there's little risk of the ambiguity that often exists: Did the patient avoid activity X and therefore get Alzheimer's or did the person already have Alzheimer's and therefore chose not to adopt activity X? In this particular study, the monolingual participants actually had more education and higher job status, yet they fell to Alzheimer's sooner. One alert: muddling along at a partial level of bilingualism appears not to qualify for these dramatic protective effects, meaning my "excellent" performance in my scientific Russian course at the University of Alberta was all for naught, at least with regard to dementia. (But it helped a lot being able to read Cyrillic lettering in the coverage of the 1972 Canada-Russia series from Moscow.)

It's not really obvious why being bilingual protects against dementia, but psychological studies have shown that bilinguals are more efficient in the use of various brain areas and are faster on tasks like naming, as quickly as possible, shapes or colours of shapes as they're switched back and forth.[9] These differences aren't apparent in young people, but with advancing age, monolinguals use more brain areas to accomplish the same task and still don't do it as well. So it's not so much the second language that counts but the apparent recruitment of so-called executive functions that is protective.

Another surprising influence is conscientiousness. That's right: the more goal setting you do and the more determination you have to achieve those goals, the more efficiency, organization, thoroughness, self-discipline and reliability you have, the more you demonstrate conscientiousness, an attribute easily determined by psychological tests. If you score in the ninetieth percentile on such tests, you have an 89 per cent reduced risk of Alzheimer's compared to those who score in the tenth percentile, at least according to an examination of nearly one thousand people over twelve years as part of the Religious Orders Study.[10]

Again, it seems odd to associate conscientiousness with the onset of Alzheimer's, but it does parallel educational attainment and is also linked to resilience, both of which are likely also playing a defensive role against Alzheimer's. On the other side of the coin, a recent study out of Finland argues that cynics are more likely to become demented.[11] Out of a total of 622 people, 46 became demented over the course of the study, and "cynical distrust" stood out from a number of other factors as being associated with dementia. Connect that to the failure of the brain if you can.

It's really important to be on guard against effects that are correlated, even though they might not appear that way. So for instance, a challenging occupation will more often fall to someone with a good education, so the two are not separable. And while it might appear that a challenging occupation is building brain power, an unskilled one might bring a person into an unhealthy workplace, a situation that might introduce its own influences. One study went so far as to suggest that a father's occupation, skilled or unskilled, and a family size greater than seven could influence the risk of Alzheimer's among the children.[12] The father's occupation is obviously an indirect effect, but the authors suspected it could influence income, nutrition and even adequate medical care. It was somewhat surprising that family size could influence the risk of eventual Alzheimer's disease, but a large family could contribute to risk of infections that would somehow predispose a young brain to eventual dementia. The study was critiqued in the very same issue of the journal in which it appeared as possibly flawed by a biased selection of cases (those who chose not to participate might not have been a random sample) and narrowness of approach (socioeconomic factors act not just in childhood but over a lifetime). Nonetheless, it demonstrated how far afield researchers are willing to look for possible influences.

Leisure activities are a fascinating set of influences because it's obvious that they represent a broad range of mental stimulation, from crossword puzzles (good) to television (not so good). I feel it's my duty, having worked in television for years, to bring one particularly damning examination of the effects of TV watching to light.[13] Participants in this study were in their seventies and eighties and already had Alzheimer's disease, but a list of their leisure habits in the years before, when they were between forty and fifty-nine years old, were collected from friends and family members. Those activities fell into a broad range of categories: social (going to church, talking on the phone), intellectual (playing an instrument or board games) and physical (skating, walking or swimming). The Alzheimer's patients had watched more TV than the controls—27 per cent of all their daily leisure activity hours versus 18 per cent for those who did not have the disease. Controls spent an extra half hour every day on social and intellectual activities than did the cases. Each additional hour of TV per day raised the risk of Alzheimer's measurably; every hour spent on intellectual activities lowered it. And given that this study was completed in 2004 and that many of those involved spent the first thirty or more years of their lives *before* TV, the companion study in 2035 will be really interesting, since it will be made up of people who have never been without TV (if indeed there is TV then!).

The conclusion from the TV study is that some leisure activities are beneficial and some are not: crosswords are better than the tube. Such conclusions must be made with care, however. As I mentioned earlier, some studies have shown that it is the frequency of leisure activities, not the kind, that is important. In addition, at least one study listed television as a positive activity that involved "information processing," with the implication that it counted toward cognitive reserve.[14]

Note too that many leisure activities involve socializing and that having an active social network has been shown many times to enhance cognition over time. In that sense, even TV watching—in a group—would be beneficial, and it would be better than watching the tube solo.

One possible confound that researchers do their best to allow for is this: people who are just entering into the early phases of dementia, perhaps not yet even diagnosed, may choose to avoid mentally challenging activities. They might then be mistakenly allocated to the group of people who, because they avoided such activities, eventually developed Alzheimer's, when in fact it was the other way around. The value of one spare-time activity over another is, of course, an issue that is front and centre for those who are approaching their sixties, a time when Alzheimer's rates start to rise. It's one thing to say that education and cognitive activity are important when people are young, but there isn't as much evidence that benefits accrue when such activities are started later in life. However, two very recent studies suggest that yes, it's probably good to be mentally active at any age. One of these studies showed that an active mind delayed the onset of cognitive decline; the other showed that exercising one's mind also seems to reduce the accumulation of amyloid beta. It's all pretty convincing.[15]

I have been selective in the studies I've chosen to mention, and you can imagine that even more would bring more detail, nuance and complication to the picture. The kinds of tests used, the makeup of the subject group, the reliability of data—especially testimony—and the robustness of statistical relationships all combine to make it difficult to come to hard and fast conclusions. But the bottom line seems to be that mentally stimulating activities, especially at a young age but also possibly throughout life, guard against Alzheimer's disease.

But how? What actual, concrete effect do such activities have on the brain? The human brain is plastic—that is, it can develop new dendrites, new circuitry and neuronal connections. No more than a couple of decades ago, it was assumed that the human brain, once mature, was incapable of creating new neurons, that the existing neurons had lost the ability to divide. There wasn't any evidence that they had, and there was also a bias against the possibility based on the idea that the brain had to be extraordinarily stable in order to function. New connections, yes, but whole new sets of neurons, no. However, networks that are too stable have a hard time incorporating new information. And today, we know that the adult human brain can create entire new neurons through "neuro-genesis." And among the best evidence for this ability is a most unusual and clever piece of science.

From 1945 to 1963, the Soviet Union, the United States, France and Britain conducted atmospheric testing of nuclear weapons, but the Limited Test Ban Treaty, signed by all except France in 1963, forbade further testing in the atmosphere while allowing tests to continue underground. China also did not sign the treaty, but both China and France had stopped atmospheric testing by 1980. One of the components of the fallout from such tests was the carbon isotope ^{14}C, or carbon-14.

The point here is that plants absorb CO_2 through photosynthesis and eventually that carbon works its way through herbivores to carnivores to humans. During the years of atmospheric nuclear testing, increased amounts of ^{14}C from those explosions worked their way into humans and were incorporated into our cells, even the DNA in the chromosomes. When those cells divide, they must duplicate their DNA, and some of the ^{14}C in that DNA will there-fore be incorporated into the DNA of the newly created neurons. Which means at least some ^{14}C will still be in our brain cells today.

So in the first decade of the twenty-first century,[16] an inter-national group of scientists analyzed tissue from the hippocampus from brains of people who had died at ages ranging from nineteen to ninety-two. (Earlier research had suggested that the hippocampus was a likely place to look for new neurons.) The brains of people who were born before the 1950s had signifi-cantly higher amounts of ^{14}C than had existed in the atmosphere when these people were young, proving that since the subjects' youth, some of their hippocampal cells had taken in ^{14}C in the act of duplication. This was a beautiful piece of work in that it allowed for several conclusions. First, the oldest individual sampled was apparently still adding new neurons in the fifth decade of life. Second, that rate of addition seemed not to slow perceptibly with time. And further, scrutiny of the data revealed that not all hippocampal neurons were replaced: some areas were active, some were not. One particular spot within the hippocampus, called the "dentate gyrus," crazily replaces its neurons over decades, at a rate of about seven hundred new neurons per day! This productivity isn't enough to prevent the gradual loss of volume in the hippocampus over the years, but it surely mitigates that loss.

The fact that the dentate gyrus is the locus of (relatively) feverish generation of new neurons is titillating because that structure represents a kind of neural bottleneck. Crucial new information is forced through the dentate gyrus, and the result might be a premium on adaptability. Apparently, the dentate gyrus plays a critical role in "pattern separation" (for instance, separating today's breakfast from yesterday's from the one the day before — a kind of time coding for events that are otherwise similar).

If neurogenesis is to be considered an important player in the changing brain and in the delay or prevention of Alzheimer's, it has to be shown that new neurons actually participate in the workings of the brain. In mice, this is possibly true, as it's been shown that genetically identical mice (which have nonetheless developed different habits of exploration and territorial behaviour) have different levels of adult neurogenesis.[17]

This connection is much harder to establish in humans, but one classic example could be interpreted as confirming a link between the brain's work and the creation of new neurons.

Anyone who has ridden in a London cab knows how incredibly challenging it can be to navigate the warren of streets in that city. It's so demanding that there's a formal name for becoming a London cabbie and learning to navigate the city without GPS: The Knowledge. It can take three years or more to learn how to get around the whole city. And that period of intense learning leaves its mark on the brain—specifically, in the hippocampus. This part of the brain is ideally suited to the task, since it is devoted to both laying down new memories and commanding knowledge of surrounding space. In 2000 a group of U.K. scientists showed that experienced London cabbies had larger hippocampi than did control drivers.[18] In particular, the right rear part of the hippocampus was unusually large. The direct conclusion was that cabbies' specialized and encyclopaedic knowledge of London geography had swelled that part of the brain. However, the change had taken place at the expense of the front part of the hippocampus, which was smaller than in control brains.

The research even showed that the change in size couldn't have been the result of people with larger right hippocampi wanting to be cab drivers (although the authors allowed that this "would be fascinating in itself"). Because the size of the right rear hippocampus correlated with the number of years of experience, the researchers concluded that the hippocampus didn't start large; it grew over time.

The study didn't address the question of whether the size change was the result of the production of new neurons or simply the elaboration of neural circuitry. Regardless, however, it was a splendid demonstration of the brain's ability to respond physically to the demands placed on it. The results don't speak to the preservation of cognitive skills with aging, but they do suggest that education and mentally challenging occupations might be doing similar things in other places in the brain. Of course, the hippocampus might be unique; at this point, we just don't know.

And so the picture of the aging brain has changed dramatically over the last few years. We now know that it's dynamic—that it can be changed by a host of different experiences, both mental and physical. And we're just beginning to understand how these experiences might all play together to ensure that some people navigate old age intact and healthy, even if they have the physical signs of Alzheimer's in their brains. In fact, these findings might explain some recent surveys that could chip away at anxieties we have about dementia and the future. That's next.

CHAPTER TWELVE

Is the Epidemic Slowing?

Extreme caution is the only sensible approach to this new development, but some little-publicized reports are beginning to emerge, suggesting that in some countries and with certain patient groups, Alzheimer's disease, or more broadly, dementia, is declining. Not rising in epidemic fashion. Not showing signs of being the "plague of the twenty-first century" as I called it in the introduction to this book. Slowing instead.

Obviously, if these few studies actually represent a solid trend, demonstrating significant change, it's a hugely important finding.

The most striking investigation showing the decline of dementia was described in an article in the medical journal the *Lancet* in 2013 undertaken as a response to demands to know more about the future of dementia.[1] If the costs of caregiving were going to soar, everyone from caregivers to ministries of health to pharmaceutical companies needed to know. The authors of this study felt that there was a shortage of studies indicating what the future prevalence of dementia might be, especially among various age groups. They admit they weren't the first but suggest that earlier studies (which

I'll mention in a moment) were rather hard to interpret and difficult to pull together because of methodological differences like
inconsistencies in sampling and diagnosis and disparities in health
care available to the participants. In their view, it was impossible
to draw firm conclusions from those studies.

Between 1989 and 1994, researchers interviewed and tested
sixty-five-year-olds in six different areas in the U.K., using for
the most part identical survey methods. They then chose three
of those areas—Cambridgeshire, Newcastle and Nottingham—
to repeat the survey in 2008 to 2011. So there were two different groups of sixty-five-year-olds, sampled twenty years apart. In
each case, roughly 7,700 people were surveyed. When the numbers
in the earlier sample were extrapolated to the general population,
the number of people with dementia was estimated to be 664,000.
Simply taking into account the aging of that population, the group
estimated that the 2008–2011 survey would net some 884,000. But
no. Instead, the number, 670,000, was roughly the same as twenty
years earlier—214,000 less than expected.

It's important to note that this was dementia, not specifically
Alzheimer's disease, but more about this point shortly. In addition,
the research team was quick to identify a set of potential shortcomings, including (for reasons that weren't clear) a much greater
reluctance to take part in the second round of the survey. An 80 per
cent response rate in the first round declined to 56 per cent, and it
was difficult to know what might therefore have been missing in
that data. Why were people refusing to respond, and how would
their noninvolvement skew the results? It was even difficult to
decide what qualified as dementia in the midst of ongoing controversy over exactly what constitutes an accurate diagnosis.

As I mentioned earlier, this study followed several others
that reached similar conclusions, although they were conducted

in different ways. An American review of people enrolled in the National Long Term Care Survey (NLTCS) found a significant decline from 1982 to 1999: 5.7 per cent prevalence in 1982 but only 2.9 per cent in 1999.[2] (This was probably the most decisive of a small number of American studies run along the same lines as the *Lancet* study in the U.K.) Even though this review, like the British study, included all dementias, it wasn't particularly good news with respect to Alzheimer's specifically because almost all of that reduction seemed to have been in the so-called "mixed" dementias—those combining some aspects of Alzheimer's with circulatory issues in the brain. The gradual decline in the rate of stroke, combined with more effective treatments for that malady, was therefore suspected to account for most of the decline. So from this paper alone, it's impossible to say much of anything about the decline or rise of Alzheimer's in the period under investigation.

A similar study, covering a ten-year period (1990 to 2000) in Rotterdam showed enough of a downward trend for researchers to be satisfied that they'd observed a real phenomenon.[3] Although the study failed to reach statistical significance, the researchers' confidence could be attributed partly to the fact that MRI images revealed a decline in brain shrinkage and small blood vessel disease over the decade. However, mortality rates were higher in the 1990 cohort, implying that some participants would have died before developing dementia. This would have the effect of lowering the apparent rate of dementia in the group, making the decline that *was* seen even more dramatic. (Had they lived another ten years, they might well have been diagnosed with dementia at the time, thus raising the 2000 numbers and partly reducing the downward trend.) The researchers also point out that a higher level of awareness and reporting of dementia would have pushed the later numbers to artificially high levels. There's such intense interest in studies like this that even statistically *insignificant* numbers beg to be rationalized. Notwithstanding the marginal effects, the authors of the Rotterdam study argued that two factors might have accounted for the decline in dementia: reduced risk of vascular disease and—that staple of brain reserve theories—education.

These results were virtually replicated in a Swedish study of seventy-five-year-olds centred roughly on the period from 1990 to 2005. The prevalence of dementia was a shade under 18 per cent at both ends of the study. While that sounds as if nothing much happened, people with dementia were living longer in 2005 and thus more cases of dementia were being recorded. Pairing longer survival with consistent numbers means that there had to have been fewer *new* cases of dementia as time passed. Again, there's also the possibility that heightened awareness of dementia led to a higher level of diagnosis in the later period. And multiple factors were identified as having influenced the health of the individuals involved—some positively, some negatively. Reductions in smoking, improved physical fitness and lowered cholesterol and blood pressure levels likely conspired to lower risk, but rising levels of obesity and diabetes would have countered those.

Back to the *Lancet* study for a moment, where the prevalence of dementia in the later group was an estimated 24 per cent lower than anticipated. Education and improvements in general health and medicine were suggested as reasons for this result. The two cohorts of patients were roughly twenty years apart, and given a minimum age of sixty-five, they would therefore have been in school in the 1920s/30s versus the 1940s/50s. I'm not aware of any attempt to identify specific changes in educational practices in the U.K. over that two-decade period (which might have contributed to the ultimate reduction in dementia statistics), but that would be a fascinating tie-in. It might just be that people stayed in school longer.

IQ is often used in such studies as an indicator of brain reserve, and it's been well established that IQ scores have been rising steadily over the past few decades, a phenomenon called the "Flynn Effect" (though intelligence researcher James Flynn protests that he didn't coin the term).[4] IQ is often cited in journal articles as one of the proxies of brain reserve, so is it possible that rising IQs have played a role in reducing the incidence of dementia? Flynn himself argues that IQ scores measure several different components of intelligence and that scores have been rising because of improvements in some, but not all, of those areas. In a sense, today's world demands a different set of mental abilities, and some of these are kicking scores higher. It's not that we're that much smarter than our parents or that our children are brilliant or our grandparents were, well, really stupid; it's that we and our children score better on tests that, if they were updated, might make sense of those scores. As an example, Flynn argues that we are probably more adept at analysis but that our ancestors were superior in rote memory. And here's another explanation: today's smaller families mean more adult conversation in the home and a better chance for children to absorb that sort of information. The question is this: Do any of the areas tested by intelligence tests stand out as being protective against dementia? To my knowledge, no one has attempted to evaluate that possibility.

But what is the significance of the fact that these studies all looked at dementia and not specifically at Alzheimer's? This does dull the little glow of optimism that might have been created, because Alzheimer's is only part of the whole. And even within a group of people who carry the diagnosis "Alzheimer's," the situation is complicated. Of course, some brains are loaded with plaques and tangles, but others also have tiny infarcts, places that have been irrevocably damaged by loss of blood supply. The burst of a tiny blood vessel—the clot that blocks the flow of blood—creates an infarct. Some brains have many of these, and some have plaques and tangles as well as infarcts. There are also Lewy bodies, plaque-like aggregates of misfolded proteins called "alpha-synuclein." And sure enough, some brains have only Lewy bodies, some have Lewy bodies and plaques and tangles and . . . maybe even some small infarcts. There might even be a direct connection between vascular problems and the plaques surrounding neurons. A study published in early 2014 showed that patients who had the most hardening (or diminished flexibility) of the arteries had the most plaques and deposited more amyloid than others over the next two years.[5] One suggestion was that disrupted blood flow due to problems in the arteries failed to flush the plaque away as it normally would have. If only it could turn out to be that simple.

So the *Lancet* study and all the rest lumped pure Alzheimer's disease in with other forms of dementia, since they examined a wide spectrum of this type of ailment. It's very difficult, if not impossible, to make precise estimates of how the different kinds of dementia would break down, but here are some possible ballpark figures: 65 per cent Alzheimer's, 20 to 25 per cent infarcts and 10 to 15 per cent dementia due to Lewy bodies. But figures like this can't be precise; for instance, the prevalence of infarcts and related vascular disease is difficult to quantify, at least partly because there's such a wide range of kinds of damage, ranging from easily visible to microscopic. Here are some real results from a study done as part of the Rush Memory and Aging Project showing just how complex the picture is.[6]

One hundred and seventy-nine brains were examined, mostly from people who were in their late eighties when they died and all of whom had been labelled as demented. Of these brains, 157 did indeed have Alzheimer's pathology, but 54 of those had macroscopic infarcts *as well*, 19 others had signs of Lewy body disease and 8 individuals had all three. In the end, nearly half of all those who had been diagnosed with Alzheimer's (through tests when they were alive) actually had a mix of dementias. Less than half of those who had been diagnosed with "probably Alzheimer's" had *only* plaques and tangles. The single most important number is that pure Alzheimer's disease accounted for only a little more than half of all these cases.

Different studies have offered different estimates of the relative abundance of categories of dementia. Percentages changed even from one study to the next in the Rush Memory and Aging Project, where everything else was held constant. Bearing that variability in mind and looking at the Rush results crudely, you could make the simple argument that a 20 per cent reduction in the incidence of dementia would mean about a 10 per cent reduction in the number of new cases of Alzheimer's. Obviously, that would still be a good thing. But there's also widespread agreement that the most important influence leading to this decline in dementia has been better blood vessel health. That's the result of declining smoking, better fitness and better care of high blood pressure and cholesterol. So where would improved cardiovascular health exert its biggest impact? On the dementia caused by infarcts, not plaques and tangles. So that 20 per cent reduction might actually represent 15 per cent vascular dementia and 5 per cent Alzheimer's.

That guesstimate of a 5 per cent drop in new Alzheimer's cases over twenty years could be much too low, but even at 5 per cent, the numbers are stunning: they reverse what appears to have been a stupefying upward climb. The public health implications would be profound: reducing the annual numbers of new patients would have a huge cumulative effect over time. But optimism is hard won in this business, and studies like this are rarely seen as the answer. They're just the beginning: more studies will be needed, with longer runs and more detailed examinations of the brains after death.

Still, these results have important implications right now. The fact that a group of patients diagnosed with Alzheimer's disease turns out to be a much more heterogeneous group than expected adds one more element to the uncertainty of things. Yes, plaques and tangles hold their place as the accepted diagnostic indicators, as they have for a hundred years. But there are all the doubts about their respective roles that I've already reviewed: plaques don't track the spread of the disease, but tangles do. Tangles aren't unambiguously causative either, and both might represent the end of the disease process, not the beginning.

The revelations that a significant fraction of dementias appear to have been prevented due to improvements in cardiovascular care have moved the role of the brain's blood supply back into the spotlight. Remember that throughout much of the twentieth century, the brain's failing blood supply was assumed to underlie dementia. Then, with the ascendance of Alzheimer's, vascular issues were set aside. Yet they persist and can account for a significant number of cases of dementia, some estimates arguing that they're found in 30 per cent of all cases called Alzheimer's.

This uncertainty, while not at all unusual in medical science, has raised serious questions about the design and development of anti-Alzheimer's medications. Until now, they have been largely, if not almost entirely, aimed at plaques—either preventing their formation or breaking them or eliminating them after they've been deposited. It's worth asking whether the less-than-complete confidence in the importance of plaques justifies the vast majority of funding directed at them.

And there are more naysayers in the picture. One of the most prominent is the late Mark Smith, whose prose, as much as his arguments, has made him worth listening to. Anyone who is lead author on a paper titled "Copernicus Revisited: Amyloid Beta in Alzheimer's Disease" is bound to startle the research community. In that 2001 report, he and his coauthors argued that Alzheimer's research was in a similar state to that of pre-Copernican astronomy, where the earth was considered to be the centre of the solar system. Earth = amyloid. It wasn't just the typical bait-and-switch, where the wit is limited to the title and not the content of the paper. Smith et al. loved playing with the idea:

> *This review is being submitted as a rather Lutherian attempt to "nail an alternative thesis" to the gate of the Church of the Holy Amyloid to open its doors to the idea that aging is the most pervasive element in this disease and A beta is merely one of the planets.*[7]

Elsewhere in the article, Smith and his colleagues admitted that had this whole issue played out in the ivory towers of academia, it would have had little impact on the real world, but they pointed out that this was far from the case. They contended that despite the millions of dollars poured into amyloid research, little had been learned about plaque deposition and aggregation in the living brain, pre-mortem diagnosis or any effective treatments. Their main points were that aging itself appears to play a more central role in Alzheimer's than does amyloid and that amyloid, while likely a player, is just that: a player, not the centre of the Alzheimer's universe.

They bolstered their argument by reviewing several other non-amyloid processes that seem to be involved in Alzheimer's, such as oxidative stress, long-term injury and inflammation. But Smith still wasn't finished. In a 2009 article, he (and others) used the two well-known anomalies of plaques (the fact that they don't correlate well with the spread of dementia in the brain and the fact that many people have them with no accompanying mental deficits) to advance the surely heretical idea that plaques, rather than being the best target for anti-Alzheimer's medications, were actually generated by the brain to corral the real bad actors, the small precursor molecules.[8] These toxic versions tend to aggregate into plaques, and Smith's group pointed to other biological situations where aggregation is actually protective, taking the most virulent kinds of molecules out of circulation. The researchers then wondered whether this might also be true of amyloid. In that case, plaque busting or attempts to prevent plaque from forming could be seriously counterproductive.

It's fair to say that the idea of plaques playing a protective role is not Smith's alone. Even Dennis Selkoe has suggested that this might be the case. Selkoe is a prominent supporter of the dominant scenario, the "amyloid cascade" (the idea that the accumulation of amyloid leads to plaques, tangles, synapse destruction and cell death, eventually overwhelming the brain). But he agrees that plaques might protect in the sense that they mop up the short pre-plaque aggregates and prevent them from causing damage.

According to this view, plaques don't cause any damage themselves; the toxic versions are the much smaller clumps that collect to form the plaques. These clumps might hold half a dozen molecules of amyloid beta, whereas a plaque has more than a million. But the presence of plaques could be dangerous anyway: even if they are holding onto those clumps, at any moment some are being captured but some are escaping, and that leakage is hazardous.

Smith's argument that too much weight has been put on amyloid has attracted a great deal of attention—both pro and, like the leakage counter-argument, con. And there's still more evidence for amyloid as a key player, some of the best coming from genetic studies of those who, like Auguste D., fell prey to the early-onset version of the disease. Most researchers in the field would argue that the much more common late-onset version is the same disease, or at least very closely related, so it makes sense to pursue research about the plaques that appear in early-onset. But maybe the two aren't so closely related, and what's true of early-onset may be irrelevant to late-onset. Along those lines, some have argued that Alzheimer's could be like diabetes, where type 1 and type 2 share characteristics but are triggered by different mechanisms.

If you gather all the contrary evidence, you may even arrive at a view that there's no justification for separating the normal process of aging from Alzheimer's at all. Peter Whitehouse, a prominent and well-respected, though somewhat anti-establishment, Alzheimer's researcher, has written: "In some sense we would all get Alzheimer's if we live long enough."[9]

Whitehouse is arguing that there is likely more harm than good in diagnosing Alzheimer's when the diagnosis, at least when compared to that of other illnesses, is very slippery. So this calls for a new approach. If you compare the blow delivered by the diagnosis to the inadequate options for treatment, it's hard to argue with that point of view. At the same time, the majority of researchers remain loyal to Robert Katzman's famed statement of 1976: ". . . senile as well as presenile forms of Alzheimer are a single disease, a disease whose etiology must be determined, whose course must be aborted, and ultimately a disease to be prevented."[10] And it's true that at least at the moment, from the point of view of good candidates for possible treatments, there's no viable alternative to targeting amyloid beta plaques. But only time will tell whether this kind of research has been worth so much effort.

And so the stage is set to consider the tightly linked twin topics of genetics and treatments. They cannot be separated because it's the genetics that provides the underpinning for thinking of amyloid as target number 1 for treatment. And obviously, treatment is the topic that most of us are interested in, in the end. We want to know the answers to these questions: "Am I likely to get it?" and "If I do, what will help me?" Sadly, as is all too common in leading-edge medical science, the answers available today won't satisfy us. The genetics underlying Alzheimer's have been only partly uncovered and can be wildly confusing, and today's treatments have been barely effective — at least so far. But this is where the action is: research and development. Where is the first really great anti-Alzheimer's drug going to come from, and when?

It's a risky business, this focus on amyloid — a tightrope walk of science with enormous political and financial pressure behind it. And so far it's one that, at least in the eyes of some, has prevented the exploration of other, much less well attested ideas. Nobody has put it more starkly than Mark Smith: "Realistically speaking, the perversion of the scientific method, and manipulation of a desperate public afflicted by an expanding, devastating, and incurable disease, characterize AD research and treatment in the 21st century."[11]

Strong words surely, likely too strong for most researchers, but Smith was giving shape to misgivings that many feel. And the issue is much larger than the language used, for if treatments are to come, the targets of those treatments must be accurately identified.

CHAPTER THIRTEEN

Am I going to get it? And if so, when?

Sometimes the simplest, yet most important questions are the hardest to answer. As a result, science often fails to give us the certainty we crave. With respect to Alzheimer's, the best that can be done right now is to provide tentative, often unsatisfactory, versions of what the ultimate answers to the most important questions might be.

Those queries in the chapter title demand a deeper knowledge of dementia than anyone has at the moment, but if there is a science that will be able to answer those questions, it is genetics. However, it's not the genetics of cross-breeding sheep or why you have blue eyes but your brother's are brown; it's *molecular* genetics. How DNA exerts its influence and what can go astray in the process. This is where you have to turn if you want to get a grasp on dementia and, more specifically, Alzheimer's disease.

In Alois Alzheimer's time, there was little or no recognition of the fact that genetics—family history—might play a role in dementia. As I said earlier, there wasn't even consensus about the existence of genes. The central role of genetics in understanding

Alzheimer's disease came from an unexpected direction: studies of people with Down syndrome.

In the late eighteen hundreds, life expectancy for those with Down syndrome was twelve to fifteen years. Some managed to live a few decades, though, and doctors began to notice that they seemed to exhibit symptoms of dementia. But it really wasn't until the late 1940s that the connection between Down and Alzheimer's was made precise. That's when George Jervis reported the results of brain autopsies of three Down syndrome individuals in their late thirties and early forties, remarking on the combination of "mental deterioration" with the characteristic Alzheimer's pathology of "degeneration of neurons, numerous senile plaques and Alzheimer's neurofibrillary tangles."[1]

Jervis noted that with the exception of the early age of onset, these cases seemed in every way to be typical cases of senile dementia. It's difficult to see how this correspondence could have been anything more than a curiosity in 1948, given the general acceptance that dementia was an inevitable companion of aging, not a disease. But in fact the Down-Alzheimer's connection was about to fling open the doors to new ways of thinking about both.

Ten years after Jervis, in 1958, French geneticist Jérôme Lejeune announced that he had discovered that Down syndrome is the consequence of having an extra copy of chromosome 21, three instead of two, a condition called "trisomy 21." Sometimes it's the entire chromosome that is extra, sometimes only a fragment. Lejeune was following up on work on the unusual fingerprint and palm crease patterns that people with Down syndrome have. Since he knew that these are established in fetal life, he suspected that Down syndrome might be an inherited condition already at work in the womb. Only two years before, scientists had established that human cells contain 46 chromosomes (after decades of the mistaken belief that there were 48); Lejeune exhibited photographs showing that in Down syndrome, there were 47.

There was one slight problem: he hadn't done the work himself. This was an out-and-out case of stealing credit. The real story is this: Lejeune's supervisor, Raymond Turpin, had suggested years before that Down syndrome might have a chromosomal cause. The adjustment of the human chromosome number from 48 to 46 apparently prompted Turpin to complain that no one was following up on his suggestion, so a young woman, Marthe Gautier, recently returned from Harvard, offered to take on the task of testing his idea. Despite working with totally inadequate equipment, she showed that tissue from Down syndrome indeed contained 47 chromosomes, but her microscope wasn't good enough to take science journal–quality pictures. Enter Lejeune. A frequent visitor to her lab, he offered to take the microscope slides and have proper pictures taken. Exit Lejeune. Next thing you know, in August 1958, he announces to the International Conference of Human Genetics in Montreal *his* discovery that Down syndrome is caused by an extra chromosome![2]

Gautier was the second author on the paper, but she had been robbed of the credit for the discovery. The controversy flared up again in 2014 when an awards ceremony in January 31, 2014, in Bordeaux, which was to honour Gautier for the discovery of the cause of Down syndrome, was blocked. By whom? The Fondation Jérôme-Lejeune. [3]

That extra chromosome, and its several hundred genes, disturbed normal development in a number of ways. Happily, Down syndrome patients live much longer today than they used to, but that fact has only emphasized the link to Alzheimer's. Some estimates are that anyone with Down syndrome who lives to sixty-five will inevitably have a brain full of plaques and tangles, and three-quarters of such people will have the full mental effects of dementia. They get the plaques and tangles and dementia much earlier, but just as in the general population, not everyone with Down syndrome gets Alzheimer's.

Making this connection between an extra chromosome and an accompanying high risk of plaques and tangles puts a new and much sharper focus on the genetics of Alzheimer's. It is true that there had been earlier hints of a genetic link: reports had occasionally appeared back in the 1920s and 1930s, identifying families in which Alzheimer's appeared with greater than normal frequency. There were, for instance, sporadic reports of identical twins who got Alzheimer's together, but seldom were the brains autopsied to look for evidence of plaques, tangles and cell death, so the reports languished.

It wasn't until the late 1970s that Leonard Heston at the University of Minnesota suggested that the dementia of Down syndrome, the third chromosome 21 and Alzheimer's disease all had something in common. Heston had analyzed a group of Alzheimer's patients and had found an unusually large number of people with Down syndrome among their relatives: six, where one might have been expected. So he concluded that somewhere out of the mix of plaques and tangles and extra chromosomes could come clues, not just to the effects of Down syndrome, but to a genetic foundation for Alzheimer's disease.[4]

A genetic foundation for Alzheimer's disease. Heston couldn't have imagined how far the science of genetics would come and how those words would one day be interpreted. In a significant way, the genetics underpinning Alzheimer's disease really is the *science* of Alzheimer's. As the focus tightens—from the demented person to their brain, neurons and then molecules within those neurons—you find yourself in the genetic world, where DNA, RNA, proteins—molecules all—are the actors.

The idea that chromosome 21 might have something to do with Alzheimer's was at the time intriguing, maybe even startling, but there wasn't much that could actually be done about it. The avalanche of technology that was to revolutionize not just what we do with these molecules but also how we think about them was yet to come.

But now, with twenty-first-century eyes, let's look at the genetics of Alzheimer's disease. From the Down syndrome evidence, it looks as if there could be an Alzheimer's gene—or Alzheimer's genes—on chromosome number 21. What does that mean?

Chromosomes are packages of DNA, but they are visible in the microscope only when cells are about to divide. In the intermission between divisions, the DNA unravels into a much more diffuse state called "chromatin." But it's fortunate that the DNA aggregates periodically. This has greatly simplified the task of mapping the locations of genes in relation to each other. (However, "simplified" was hardly the right term in the beginning, decades ago, when years would pass between one crucial discovery and its follow-up. Chromosome mapping has been actively pursued since the early days of the twentieth century, and though the technologies have been revolutionized more than once, the job is still incomplete.)

It began with fruit fly geneticists realizing that certain genes were found on the fly's sex chromosomes and so could be tracked through generations of males and females. Another sixty years and the visual inspection of changes in chromosomes made it possible to pin down genes to other, non-sex chromosomes and to illustrate how the closer a gene is to others on the same chromosome, the more tightly they ride together through generation after generation of matings, surviving the splits and rearrangements common in reproduction. Then, much more recently, it became possible to identify a gene's individual subunits (the nucleotides) in order—thousands and thousands of them that cannot be seen but can be identified chemically. This development led directly to the science of genomics—gene identification and mapping taken to levels that would surely boggle the minds of geneticists of even a generation ago.

The sequences of subunits on the DNA are translated by machinery in the cell to comparable sequences of subunits (amino acids) in proteins. Proteins do almost all the work in living cells, and their ability to fulfill that role is predicated on their having been assembled accurately, piece by piece. Any disruption of the DNA-to-protein translation process is potentially harmful.

The translation code is a 3:1 affair—three consecutive nucleotide subunits in DNA specify one particular amino acid in a protein. Change one of those nucleotides and, more often than not, the altered code calls for a different amino acid. It may even stop the growing chain. Add or subtract a nucleotide, and the entire reading frame shifts, just as dropping a single letter into a set of three-letter words changes all the ones that follow. These changes and substitutions are mutations, and their impact isn't so much on the DNA where they reside but on the mutated proteins that they create.[5] *

With this knowledge of how DNA creates proteins, we now have a much better idea of what's happening on chromosome 21. One gene there is responsible for the production of a protein called

---

* I don't want to wander too far from the DNA-protein relationship, but it has to be said that much of our DNA doesn't actually bother with all this. A mere 2 per cent is dedicated to protein production, while the rest is involved in regulating the 2 per cent or ensuring that the entire, vast mechanism is chugging along properly. This 98 per cent used to be called junk DNA, but I never actually met scientists back then who claimed that junk DNA didn't do anything. They just didn't know what it was doing.

"APP" (the amyloid precursor protein). I mentioned it in Chapter 8—a long, sinuous protein, whose function isn't fully understood. It is known to be processed in different ways, yielding different-sized products. In neurons, it's transported along the axon to the synapse, the communication portal to adjacent neurons, where it's inserted into the surface membrane of the cell, snaking in and out of the membrane. It doesn't remain intact for long, though. Enzymes cut it to pieces, and each piece assumes a unique role. For everything to work as it should, these little acts of enzymatic surgery have to be performed at exactly the right place. And that place is determined by certain contours or clefts in the enzyme molecule having an affinity for the mirror-image locations on the amyloid precursor protein. Hand-in-glove, lock-in-key, only much more complex and dynamic because the lock and the key are both trembling and twisting, and the crowd of molecules around them is jostling them. It's a tight spot, demanding a high degree of accuracy.

Any mutation, any substitution of one amino acid for another, any disarray introduced into the mix is likely to cause trouble. And so it does. By now, more than thirty mutations have been discovered along the length of the amyloid precursor protein, many of them right in the target site of one of the enzymes that slice this, the parent molecule, in two. That misfit causes inaccurate dissections of the amyloid precursor protein, and this, in turn, leads to the release of amyloid beta, the bad actor in Alzheimer's disease, and specifically the version that is forty-two amino acids long. That version is fond of aggregating with others like it, eventually to be deposited as plaques in the brain.

But it's not just inaccurate dissection of APP that is disastrous. Producing too much of the protein has the same effect. This is what happens in Down syndrome, where a complete replica of the amyloid precursor protein sits there on the extra chromosome 21.

So now there's an explanation of why people with Down syndrome inevitably accumulate loads of plaques in their brains if they live long enough. It's a satisfying resolution of a puzzle, but sadly, the errors in the expression of this APP gene solve only a tiny part of the Alzheimer's genetic story. For one thing, chromosome 21 plays a role primarily in early-onset, familial Alzheimer's, the version that runs in families and strikes early, usually between the ages of fifty and sixty (but sometimes much earlier). And even then, only about 10 or 15 per cent of those familial cases are caused by these mutations.[6]

And the returns continue to diminish: familial cases represent only about 50 per cent of all early-onset Alzheimer's, and early-onset comprises no more than about 5 to 10 per cent of all Alzheimer's. If we add it all up, chromosome 21 defects amount to—*at most*—about 0.5 per cent of all cases of Alzheimer's dementia.[7] But this line of research is still extremely significant. It deserves notice because it establishes certainty that genetic mutations can lead to Alzheimer's. And some extended families where such mutations are rampant provide unique opportunities to test ideas for new treatments—more on that in Chapter 14.

Now to the question "Will I get it?" Based on this research, the answer is tricky. Some of the mutations that cause early-onset familial Alzheimer's disease are dominant, meaning that you need to inherit only *one* copy of the gene to be sure to get the disease. Not all mutations are so deterministic, though, and the decision about whether or not to seek genetic counselling will depend on how anxious a potentially susceptible person might be. You might not want to know that you'll inevitably be struck down by an untreatable and undeniably unpleasant disease, but many people have good reasons for wanting to know and, according to some studies, can deal with the worst-case scenario surprisingly well.[8]

One of the most exciting recent discoveries involving APP focused on a large cohort of aging Icelanders. DeCODE genetics, a subsidiary of the biotechnology company AmGen, reported in 2012 that it had found a mutation in the APP gene that, contrary to virtually all others, actually *lessened* the risk of Alzheimer's. This mutated gene was much more common in eighty-five-year-olds who were cognitively intact than in their demented counterparts. In fact, those Icelanders who carried the gene were consistently sharper at every age.[9]

The announcement gave the Alzheimer's world a little positive jolt, not so much because it was nice to hear about a bunch of intellectually sound Icelandic elders, but because it encouraged those who were designing drugs against plaque based on the amyloid cascade hypothesis. This theory has its detractors and it's not airtight, but when a gene mutation is discovered that affects the way the amyloid precursor protein is sliced up—and lowers amyloid beta as a result—further investigation could be a rewarding route to treatment. (DeCODE has been at the centre of controversy for years. To build the database of genomic information, it needed access to Icelanders' personal health records, and while the Icelandic government gave deCODE that permission, invasion of privacy claims and poor financial performance soured many Icelanders on the company. DeCODE was finally bought by the American company AmGen in 2012, and much of the controversy has receded.)

DeCODE's discovery of a beneficial mutation in the amyloid precursor protein APP bordered on sensational, but just as important, at least to biochemists, is that the discovery underlined just how important this particular site is in the APP gene. Why? Because a different mutation in precisely that location, already familiar to scientists, increases the risk of Alzheimer's. Code for one amino acid, risk goes down; code for another, risk goes up.

Two other genes besides the APP gene are prominent in the story of early-onset Alzheimer's—called "presenilin 1" and "presenilin 2." The remarkable rediscovery of Alois Alzheimer's slides of Auguste Deter's brain that I described in Chapter 2 perfectly captures the impact of one of those genes, presenilin 1. Remember that Auguste was already well on her way to dementia at the age of fifty-one, when she first saw Alzheimer. So she was obviously suffering from the early-onset version of the disease. When a slice of her brain was analyzed genetically after Alzheimer's slides were unearthed in the 1990s, it turned out that she had a mutation in the gene presenilin 1 (a mutation not found in any other patient so far, one that changed a single amino acid out of more than four hundred). Unique, yes, but in a location very close to other mutations that have been discovered. Two of these other mutations have been found in different specimens right next to the location where Auguste's mutation occurred, suggesting that this segment of the presenilin 1 gene is very important—and vulnerable. So presenilin 1 seems to be an exceptionally mutation-susceptible gene: as I'm writing this, 197 mutations (!) have been found in it.*

The presenilin 1 gene sits on chromosome 14 and is part of an enzyme complex involved in slicing up the amyloid precursor protein. Presenilin 2 is the third gene (besides APP and presenilin 1) that's implicated in familial early-onset Alzheimer's. It participates in exactly the same process, even though it's on chromosome 1, the other side of the genomic planet from both APP and presenilin 1. So clearly, we're seeing that it's not really significant on which chromosome these genes reside, but rather what they do. All are involved in one way or the other in chopping the amyloid precur-

---

* Again, it must be acknowledged that a subsequent attempt to confirm Auguste's presenelin gene failed.

sor protein into pieces. Despite their similar actions and names, however, presenilin 1 and 2 make very different contributions to Alzheimer's. Presenilin 1, with its myriad mutations, is a significant player, responsible for as much as 75 per cent of early-onset familial Alzheimer's. Presenilin 2 verges on insignificance, accounting for only 5 to 10 per cent of cases.

Taken together, however, these three genes and their mutations account for only about half of all early-onset *familial* Alzheimer's disease; so many families that encounter early-onset carry none of the three. And the early-onset familial version of the disease is but a small part of all Alzheimer's disease—maybe 1 or 2 per cent.[10] All the same, their significance outweighs their rarity. For one thing, the fact that they are all involved in triggering disease by messing up the proper dissection of the amyloid precursor protein lends weight and credibility to the amyloid cascade hypothesis. Regardless of the observation that many people can have a brain full of plaques and yet be fine mentally, these mutations are saying, "Too much amyloid beta plays a crucial role in creating Alzheimer's." And that fact opens the doors to possible treatments.

But up to this point, although the challenge of finding and understanding APP and the two presenilin genes has inspired some really fine science, we are just scratching the surface of the global problem of Alzheimer's. These three genes (and likely more to come) are significant because they *cause* Alzheimer's disease. However, the most important genetics, or at least the genetics of greatest significance to a much larger number of people, is the science of the genes that *predispose* to Alzheimer's. Genes that put one at risk. And there is one that stands out. It's called "apolipoprotein E" or more usually "APOE" (pronounced "A-Po-EE").

APOE is a fascinating gene. It has a curious history, it builds proteins that do many things and it comes in three versions (each having a different influence on Alzheimer's). Unlike the three genes we were just talking about, which determine early-onset Alzheimer's, APOE predisposes people to the late-onset version (which begins after the age of sixty-five and often not until eighty-five). Underlining that idea is the recurring caution in the scientific literature that "APOE is neither necessary nor sufficient" to cause Alzheimer's.[11] Instead, as several authors seem to agree, it tweaks both risk and the age of onset of Alzheimer's, either accelerating or delaying it, depending on which version of the gene the person has.

The three versions of the gene are APOE2, APOE3 and APOE4; you inherit just two, but they come in six different combinations. Any appearance of APOE2, let alone a paired appearance, is rare: only about 7 per cent of the population has it. But this group is very fortunate because this particular version of the APOE gene *lowers* the risk of Alzheimer's—not dramatically but still significantly. APOE3 is by far the most common version, amounting to 75 per cent of the total, and it seems to have little impact on risk. However, those with APOE4 have a significant, even dramatically heightened, risk. It's been estimated that one copy raises the risk of Alzheimer's three- to fourfold; two copies, as much as fifteenfold. But here's where we come to the fact that this gene is neither necessary nor sufficient. Even with the shocking boost in risk generated by APOE4, a third or more of all Alzheimer's patients have no E4 genes at all, and as many as 50 per cent of those with two copies of the gene live to eighty and are not demented![12]

There are other combinations too: APOE3/4 tilts toward Alzheimer's, as you'd expect with APOE4 as part of it, but the added risk happens only in women. Men who are APOE3/4 have no greater risk than average. More recent data have extended the idea of women being more vulnerable to the gene: having one copy of APOE4 was of little or no additional risk to men, but it represented a significantly heightened risk to women.[13]

This is a reminder of the fact that late-onset Alzheimer's is malleable: diet, exercise, education, a mentally challenging occupation, conscientiousness and likely many more as-yet-unidentified influences play a role in creating the brain reserve that is considered to protect against these genetic effects.

Even so, the anxiety created by possession of the APOE4 version of this gene is real. As I mentioned before, James Watson (co-discoverer of the structure of DNA) had those genes redacted when his personal genome was made public. Seventy-nine at the time, Watson didn't want to know his genetic status, especially because his grandmother had had Alzheimer's disease. However, genome scientists pointed out that the rapidly growing catalogue of nearest-neighbour genes would actually allow them to determine Watson's risk status even if his specific set of APOE genes were blacked out (they merely pointed out the possibility—they didn't violate his desire for privacy).[14]

The APOE gene sits on chromosome 19 and exerts its effects in myriad ways. It influences the inflammatory response to infection, it is intimately involved with cholesterol (moving it in and out of cells, not just in the brain but all over the body) and the APOE4 version can cause abnormal brain blood flow and slower recovery from head trauma in teens and young adults (and head trauma is a predisposing factor for Alzheimer's). People with APOE4 build up high levels of circulating cholesterol, get hit by vascular disease *and* accumulate amyloid beta plaques in the brain. The E4 version of apolipoprotein is also somehow more likely to permit the buildup of amyloid beta. But it's all a bit murky: it's not known exactly what prompts mutations in APOE to trigger plaque formation. That discovery, if it's ever made, will take geneticists down a long and mysterious pathway.

This is just another chapter in an oft-told story in genetics: tiny substitutions in a piece of DNA create new versions of genes (in this case, three different APOE genes), which, when the DNA sequences are translated into proteins, create dramatic downstream effects. Somehow those changes alter the APOE protein's preferences in bonding to other molecules, with the result that plaque deposition is either ramped up or not.

How did it all end up this way? Why would we even carry a gene that is potentially so destructive? The answer lies in APOE's history, which is a curious one. For one thing, this gene is distributed widely among other species, including our closest relatives, the chimps. They have just one version (not the three that we have), and the amino acids in their proteins are different, but the chimp gene most resembles our APOE4.

Caleb Finch, a specialist in human aging at the University of Southern California, has argued that the rise, in humans, of APOE4 (which, as just mentioned, is not exactly the same as the chimp version) was the result of our transition away from the chimp line.[15] As we evolved, we began to eat much more meat than chimps do, and that is still true. That change in diet precipitated a new series of risks: hunting accidents, bacterial and viral infections from inadequately cooked meat and zoonoses (diseases that jump from animals to humans). Heightened defences against infection would have allowed the survival and reproduction of those equipped with them, and one of those key defences is inflammation, which slows the spread of infection and afterward participates in wound healing. APOE4 facilitates inflammation in some circumstances, and even its role in transporting and storing fat would have been helpful in an era of intermittent, unpredictable food supply. In Finch's words, APOE is a "meat-adaptive gene."

Those carrying the human APOE4 gene would have benefited from it in hunter-gatherer days, but today it's a different story because many more of us live into old age, when the cumulative effects of APOE4 turn around and bite us. Finch argues that this factor might explain the gradual evolution of the different forms of APOE in humans, with APOE3 becoming by far the most common form of the gene. It appears that this version arrived only something like two hundred thousand years ago, millions of years after the chimp-human split but before the large-scale movement of modern humans out of Africa. And while it performs the same roles as APOE4, it does so with moderation. In an infection-saturated environment, APOE4 would be better, but once one has survived that and acquired immunity to some of the dangerous bugs, APOE4 becomes a handicap. It might be that the very long-term good news is that given APOE3's relatively rapid rise in the human genome from two hundred thousand years ago to today, this version might eventually eliminate both APOE2 and APOE4—though long after all of us have died.

For years, APOE4 has been the only gene geneticists could connect directly to late-onset Alzheimer's. Even after several genome-wide studies had been completed, it was the only reliable Alzheimer's-linked gene. But now there are at least five more candidates, a direct result of one of the most amazing transformational developments ever to occur in a scientific industry—the industry in this case being high-volume gene sequencing. The top three candidates have names that can be found only in genomics: CLU (clusterin), CR1 (complement component receptor 1) and PICALM (phosphatidylinositol–binding clathrin assembly protein).

I've concentrated here on the genetics of Alzheimer's, but the title of this chapter is, "Am I going to get it?" and the answer includes more than just genes. There's brain reserve, of course, but also the hazards of life might raise the risks somewhat, things like traumatic blows to the head, chronic sleep problems or even general anaesthesia.[16] That short list is by no means complete, and we have yet to learn the identity of all the factors that influence the risk of getting Alzheimer's. But there's no doubt that genetics is going to be central to the search for treatments.

More extensive probing of the human genome, in finer detail, in association with the presence of Alzheimer's disease, will reveal more genes that have an influence on the malady. Likely *many* more, but the influence of each could be slight, and the emerging picture is one of a vast number of genes with potential for contributing to the disease but with no single one standing out among the rest. This multiplicity hugely complicates the push to develop treatments.

CHAPTER FOURTEEN

Treatment: Candidates But No Champions

We have no really effective, long-lasting treatment for Alzheimer's disease. Politicians, correctly reading the tea leaves of a future potentially catastrophic situation, are anxious to convince us all that a treatment is in the pipeline, with dates like 2025 being tossed around as a deadline for something effective. That year may seem a long way away, but in terms of drug development, especially for something like Alzheimer's, it's not. That point was underlined by the fact that the G8 countries held a dementia summit in London in December 2013 in an attempt to accelerate the process to a treatment (although when you read that the goal is "disease-modifying therapy," you start to get a hint of the immensity of the proposition).

To date, the best that's been done is to bring a small set of drugs to the market that ameliorate the situation for Alzheimer's patients for a year or less, after which (in reality, *during* which) the progression of the disease continues unaltered. I'll outline how those drugs work and why, by their very nature, their beneficial effects are time limited. Ironically, given that the history of

Alzheimer's from the very beginning has been characterized by the focus on plaques and tangles as *the* disease entities, these first drugs to hit the market addressed neither. However, the direction of research has since changed dramatically, and now it is indeed plaques and tangles—especially plaques—that are the prime targets for newly developed drugs.

But the word "target" implies something straightforward, and the search for treatment for Alzheimer's is far from that. For one thing, I've already pointed out that plaques, though generally assumed to be crucial players, have an uncertain role. They can be present in people who suffer no dementia; they locate to areas of the brain that seem not to be intimately connected to the spread of the disease; and they themselves might only be the aftermath, the "tombstones," of forebears that are the truly toxic entities. Similarly, tangles play an uncertain part because most of the evidence suggests that these aggregates of the tau protein develop secondarily to plaques, or at least to the appearance of the plaque components, amyloid beta. But their claim to fame is that they do track precisely the spread of synapse loss, neuron death and also the symptoms of the disease. While the plaques and tangles still maintain their hold on researchers' imaginations, other biological mechanisms and materials might play crucial roles in the development of Alzheimer's, including inflammation, cholesterol and insulin. There's a long and growing list of possibilities, but it's still true that plaques and tangles are right at the centre. The pursuit of treatments focused on them illustrates beautifully the complexity and difficulty of the science of interfering with processes that occur at the molecular level in the living human brain.

Of course, the brain is complex and delicate, and trying to affect a single process without inadvertently changing others is a monumental challenge, but I'd bet it's even more difficult than most of us can envision. For one thing, the brain is crowded. Most actual images of neurons select out certain kinds and so give the impression that there's a cell here and another over there. Sure, these neurons are multiply connected and, on their scale, spindly and long, but there seems to be a lot of space between them. But no, there isn't. The true picture is of cells in physical contact, cells wrapped around other cells, connections strung everywhere. There are fluid-filled spaces in the brain called "ventricles," but most brain material is pretty solid stuff. And at a microscopic level, or even the more extreme *molecular* level, that crowding never relaxes—and that's just the spatial aspect. Biochemically, things are even more crowded. The production or alteration or conjugation or removal of each functioning piece is controlled by yet more molecules. It's absurd to think that any one piece could operate on its own, independent of the crowd around it.

To the untutored human eye, the brain is complete chaos. To the dedicated biochemist, it's the complexity that makes the brain so enticing. To the Alzheimer's researcher, it's a monumental challenge. When designing a treatment, it's virtually impossible to affect only the single target in which you're interested. The side effects on other parts of the brain might more appropriately be called "collateral damage," and in this case, the potential for unanticipated knock-on effects is lurking everywhere. No matter which treatment angle you take, other systems will eventually be disrupted. How can you prevent the formation of plaques when the enzymes that produce amyloid beta are also responsible for other tasks in the cell? Subdue plaques, and you can't know what else might happen. If you engineer things in a way that under-produces amyloid beta, you might well overproduce something else. Even if the treatment you're designing appears to be explicitly targeted at, say, the formation of tangles, other pathways you haven't thought about may allow for tangles to be formed. So your treatment will have underwhelming results. That's the billion-dollar game researchers are playing.

The first anti-Alzheimer's drugs — the ones still in use today — were produced in an attempt to revive a failing brain. But that had to be done in a very specific way. The drugs operate at the synapse, the place where neurons communicate. A nerve impulse arriving at the end of neuron no. 1 triggers the release of neurotransmitters, which drift, some landing and plugging into receptors tailor-made for them on the surface of neuron no. 2. When there's enough transmitter-receptor action, an impulse fires in neuron no. 2.

The brain has many different neurotransmitters, but one that is severely depleted in Alzheimer's disease is acetylcholine. It declines because one of the first areas of the brain to be destroyed in Alzheimer's is the nucleus basalis of Meynert (named after Theodor Meynert, one of Sigmund Freud's mentors), where much of the brain's acetylcholine is manufactured. The neurons from this area project all over the cerebral cortex, and as they die, the levels of acetylcholine fall and synapses fail to function properly, directly affecting the person's memory. By the early 1980s, this much was known from autopsies. It was also clear that the problem lay not with the receptors but with the transmitters. But before making the leap to introducing drugs that might replace acetylcholine, researchers argued that disrupting acetylcholine (temporarily) in young, healthy subjects should produce the same sort of symptoms as Alzheimer's does years later. So this testing was done.

Students were administered measured doses of a memory-disturbing drug called "scopolamine."* When they were asked to

---

* In Colombia, many cases have been reported in which thieves facilitated robberies by adding powdered scopolamine to a drink, but in cases like these, it's not just loss of memory that allows thieves to get away. The dose is often sufficient to knock the victim out for hours.

remember sequences of fifteen digits, they did remarkably poorly, reflecting the very similar performance of a group of elders.

This all made sense: at the synapse level, scopolamine competes with acetylcholine for access to receptors. As long as scopolamine is in the system, acetylcholine simply can't hit the same number of targets and memory is affected. Remove it, and memory returns. In Alzheimer's, the depletion of acetylcholine has the same effect, so somehow it has to be amplified. Making more acetylcholine is one obvious possibility, and this is why taking lecithin, its precursor, has been and is still recommended on a variety of websites. The problem is that the results with lecithin have been almost uniformly disappointing. For some reason, it doesn't work.

The approach that was eventually adopted and resulted in the Alzheimer's drugs we have today was to preserve the existing molecules of acetylcholine beyond their best-before date. Normally, at the synapse, transmitters like acetylcholine have only a few milliseconds to act before they're dismantled by the enzymes that hang out there. If the transmitters were allowed to float around, repeatedly plugging into receptors, neuron no. 2 would become overstimulated. So enzymes degrade the transmitters, and the components are recycled. It's a fine balance but one that goes wrong when, as in Alzheimer's, the quantity of transmitters is dramatically reduced. So the current remedies (anticholinesterase drugs) are designed to inhibit the enzymes that break down acetylcholine. The transmitters survive much longer, repeatedly plugging into receptors, giving the illusion that there are actually more of them. If the synapse normally calls for two hundred molecules of acetylcholine in order to function properly, with the Alzheimer's drugs protecting them, one hundred molecules can do the same job.

The best known of these drugs is donepezil (trade name Aricept). Three similar ones exist as well as a related drug that works on a different transmitter-receptor pair. These drugs definitely enhance memory for a limited period of time—several months—but the limitation is obvious: as more and more neurons manufacturing acetylcholine die, it becomes more and more difficult to make up for the losses by making the existing transmitters work harder. There are just not enough of them.

And that's where we stand today: these transmitter-enhancing drugs are the ones prescribed for Alzheimer's patients. They're not a long-term treatment, let alone a cure, but they can add several pleasurable months to a life.

In the Alzheimer's treatment business, however, currently available drugs are, in a sense, old news. The drugs that are in development, the ones not yet available, are the ones that could—or should—turn out to be much more important. And lo and behold, they are designed to attack plaques. Despite the uncertainties surrounding the exact role of plaques in the disease, the amyloid cascade hypothesis (the idea that it all starts with amyloid, followed by plaques, tau and tangles in roughly that order, and then death) is still widely influential. This influence has led to several attempts to develop drugs that would somehow interrupt that process. Taking a page from the history of vaccines, most of the efforts to interrupt the cascade of amyloid to Alzheimer's have been built on the idea that antibodies to plaque would be the way to go.

Antibodies represent one of the most important weapons in the body's armamentarium against disease. As in the transmitter-receptor relationship, they depend on a good molecular fit (in this case, with the viruses that invade us). A virus, especially one like the polio virus, which is consistent in its structure, is especially susceptible to antibodies. If you inject either inactivated or live but attenuated versions of the virus, the body's immune system will be alerted to the foreignness of the protein makeup of the virus and will generate antibodies to it. These are made-on-demand, novel molecules that recognize the unique shape and composition of the invader, bond to it and set in motion a variety of processes that will eliminate it. Of course, this defence happens naturally; the detailed mechanics of the process have evolved over millions of years. When creating vaccines, we just jump the gun on the natural process, as the vaccines artificially and harmlessly infect us so that our bodies will establish antibody-making blood cells. Those blood cells will then be ready to respond in case the truly dangerous target, in this case, the polio virus, appears.

That's "active" immunization. There's also "passive" immunization, whereby the antibodies are made outside the body and simply injected. Both have been tried in Alzheimer's disease, based on the following premise. As time passes, we accumulate plaques in our brains. They may not affect us negatively at all, but of course, if plaque deposition becomes a runaway process, then tangles, synapse loss, neuron death and dementia often follow in a complicated but roughly straight-line fashion. How could antibodies ameliorate that situation? There are several possibilities, but one of the best is based on the idea that it would be best to interrupt plaque formation as soon as possible—that is, introduce antibodies which will bind to amyloid beta before it aggregates. Then either brain-based immune cells called "microglia" could engulf and break down amyloid beta or antibodies could remove amyloid from the brain altogether. But sadly, the performance of drugs based on these ideas, while showing glimmers of hope, has generally been disappointing.

The basis of one such drug was a series of experiments in mice starting in the late 1990s.[1] The results of these studies strongly suggested that multiple benefits could be produced by creating an immune response to amyloid beta either actively or passively. The amount of plaque in the brain decreased, in some cases practically disappeared, and cognitive deficits were relieved. Several different labs reinforced these findings, and overall the experiments were, as one investigator called them, "tantalizing." The excitement led directly to a trial in people with Alzheimer's disease, where they were immunized using a synthetic version of what is agreed to be the most dangerous actor in all of this, the forty-two-amino-acid-long version of amyloid beta. It was called "AN1792." Of the three hundred subjects involved, some did indeed develop significant levels of antibodies to amyloid beta, and most important, in some ways this treatment worked!

Although antibody responses were, naturally, varied, there was a significant trend for those who developed the strongest antibody response to have significantly reduced levels of plaque in their brains. In two cases, almost no plaque appeared to be left at all. That was excellent news: the immunization targeted at the plaque precursor succeeded in meeting its goal of clearing plaque from the brain.

But the bad news outweighed the good. For one thing, many patients' symptoms didn't improve, even though they had much less plaque in their brains. In the end, they were still as demented as the control group. Much worse, however, was the fact that eighteen people (6 per cent of the total) developed inflammation in the brain, serious enough that the trial had to be stopped mid-run. And a handful of the eighteen suffered permanent brain damage. Although the trial was halted, patients who had been involved were followed up, and four years later, two important but contradictory findings were revealed. One was that those whose antibody response had been the best in the trial were still producing antibodies to amyloid beta and their cognitive abilities had declined less than those of the controls. On the other hand, there was no reduction in the amount of their brain shrinkage. However, it's the symptoms that matter most, so some positives came out of this experiment—enough to suggest that while this particular drug was too risky because of its potential side effects, the principle (attacking amyloid beta) was sound.

Lessons learned? Mouse brains are different from human brains or more specifically, interfering with plaque formation in a mouse is different from the parallel process in humans. Another brick in the wall of the amyloid cascade hypothesis was dislodged too: apparently, we now know that plaques can be removed without affecting symptoms. Some researchers even ventured that the amyloid hypothesis had been proven wrong, but this suggestion has not yet been widely embraced. This episode also underlined the uncertainty surrounding the introduction of any new drug: until it's actually deployed in our brains, we don't really know how it's going to turn out. At any rate, that was the end of AN1792.[2]

The AN1792 experiment was an example of active immunization: using a version of the enemy to challenge the body to manufacture antibodies to it. But researchers had also had success in mice with the alternative, passive immunization, in which the antibodies are prepared outside the body and injected. One advantage of this approach is that greater control can be exerted over the number of antibodies in play.

Again, experiments in mice showed that passive immunization seemed to be effective too, so attention turned to developing antibodies that could be injected directly into patients. One treatment, with the tongue-twisting sort of name that's common among these drugs—bapineuzumab—has gone through several trials, but again, the overall results were so disappointing that the drug has been abandoned.* With this passive antibody, there was a significant side effect: fluid buildup (edema) in the brains of several of those being treated. It caused headache, confusion and vomiting. Although the drawbacks weren't as serious as the brain inflammation caused by AN1792, they were serious enough to cause concern.

More troubling, though, given that side effects can often be managed, were the actual results. The first disappointing sign was that the patients who carried the APOE4 mutation (and so may have had the most to gain) did not improve at all relative to the controls. But in the long run, no one benefited. It's important to distinguish between effects the drug might have in the brain and impact on the Alzheimer's patient's life. With bapineuzumab, they diverged: there were definite signs of plaque reduction but no relief of symptoms or disease progression for patients. So by the end of 2012, bapineuzumab was no longer on the table.

This failure raised the same old questions: Is it really worth devoting billions of dollars to developing antiplaque drugs when they appear not to work—at least, not to work in alleviating symptoms of the disease? Or if you're a glass-half-full person, you might contend that because bapineuzumab has a greater affinity for full-blown plaques than for their smaller predecessors, it would be much more effective to target the short amyloid beta protein strings that aggregate and build plaques. According to this view, it's not that amyloid is the wrong target but that we're waiting too long to address it.

Another passive antibody has been developed, which you'd think would address that shortcoming. It's called "solanezumab."

---

* Drug names are governed by an international set of rules. In this case, the last three letters of bapineuzumab (and related antiplaque drugs) stand for "Monoclonal AntiBody.")

It has a greater affinity for the plaque precursors and so should get around the issue of arriving too late at the scene of the crime. But solanezumab, manufactured by Eli Lilly, has unfortunately had roughly the same sort of success as bapineuzumab—that is, not enough. As Dr. Derek Lowe, a prominent blogger on the topic of drug discovery put it:

> *The problem is, solanezumab hasn't shown much promise of improving the lives of actual Alzheimer's patients. Lilly's own trials showed a possible improvement in a measure of cognitive decline, but this did not show up again in a second patient group, even when they specifically modified the endpoints of the trial to look for it. And neither group showed any functional effects at all, which I think are what most Alzheimer's patients (and their family members) would really want to see.*[3]

As Lowe says, the end point of any anti-Alzheimer's drug must be a higher quality of life for the patient, and this drug, while it showed signs of being active at the molecular level in the brain, didn't provide that quality of life. However, the maker, Eli Lilly, is continuing to work with the drug, apparently basing its faith on the possibility that many of those involved in earlier trials might not actually have had significant amyloid deposits—because they were admitted on the basis of their symptoms, not on any actual evidence of plaques.

So the story on antibodies so far is that they haven't worked, and the only stray beam of light in an otherwise dark picture is that perhaps the trials haven't begun early enough, when the disease process might theoretically be most susceptible. But it's difficult to remain optimistic about this approach. A recent paper, for instance, shows that when you apply the highest technology possible to get a close-up view of what happens when these drugs meet amyloid beta, the picture is gloomy.[4] Only one of these antibodies actually seems to physically engage with amyloid beta in human tissue with any effectiveness—that's bapineuzumab. Solanezumab activity was barely detectable and another, crenezumab, appeared not to work *at all*. That's bad enough, but in mouse tissue, by contrast, they all seemed somewhat effective, highlighting once again the fact that we are not rodents.

These most recent negative experimental indications were published *after* the antibody crenezumab (the most ineffective in that study) had been recruited for what many hope will be a landmark in Alzheimer's research. This story begins with the discovery of a large cohort of relatives of Basque descent in Colombia, in which an early-onset Alzheimer's gene has been present for about three centuries. The extended family numbers about 5,000 today, of which 1,500 carry a version of the presenilin gene. So those 1,500 will get Alzheimer's disease and get it much earlier than most who will succumb to the disease. These unfortunate individuals are part of a new drug trial that started at the beginning of 2014.

It's a desperate situation, a situation worth responding to, but to be honest, this extended family is attractive from the drug discovery point of view as well. They have young and otherwise healthy brains, which lessens the confusion over exactly what might cause dementia. Moreover, the course their disease will take is remarkably predictable, making it possible to administer drugs early, *before* the onset of symptoms. In this trial, the research team can be confident that the disease *will* come.

Here is the timetable for a typical member of the family who carries the gene: memory loss begins to be noticeable at about the age of forty-five; dementia is well established by fifty. Death ensues a few years later.

As part of this trial, crenezumab is being administered.[5] It was chosen because it appears not to cause the same sort of incidental swelling or inflammation as its failed cousins, even though its mechanism of action—clearing plaque and its precursors—is the same. This characteristic implies that it can be administered safely at relatively high doses. The drug is being administered to one hundred people who carry the presenilin gene, beginning when they reach the age of thirty. They are accompanied by another hundred who carry the gene but are not being treated and a third group of one hundred who do not have the gene and, of course, are not getting the drug. Previous research has established that those carrying the gene show high levels of amyloid beta in their blood and cerebrospinal fluid and structural changes in their brains even when they are still in their early twenties. Amyloid begins to accumulate around the age of thirty, rises sharply for nine years, then plateaus. To track these early changes, the current subjects will have their brains imaged and samples of cerebrospinal fluid taken, and they will participate in other tests at regular intervals over the next few years.[6]

Beginning treatment at the age of thirty, long before overt symptoms are apparent, does sound like a radical departure from other clinical trials, where similar drugs were given to those who already had anywhere from mild to severe Alzheimer's. But given that the general belief among researchers is that cognitive changes, however subtle, are preceded by perhaps several years of amyloid buildup (and noting that imaging recorded abnormalities in the brains of these gene carriers when they were in their late twenties), it's worth wondering whether starting the administration of crenezumab even at the age of thirty will be early enough. And what if the trial fails? That would be a substantially more powerful blow to the amyloid cascade hypothesis than the failure of other drugs so far because for once, a drug is being administered early on in the disease process.

Or what if the drug works and the treated individuals are granted some sort of reprieve from Alzheimer's—anything from a few months' delay to something more dramatic? That outcome would be greeted with incredible enthusiasm, of course, but it would still leave unanswered key questions, such as this: Yes, early-onset Alzheimer's might benefit from this approach, but when the onset can't be accurately predicted (as in late-onset, the vast majority of cases), when do you start giving the medication? An irony: to apply the results of this trial to the general population would require acknowledgement that early- and late-onset forms of Alzheimer's are essentially the same, an idea that was vigorously opposed through most of the twentieth century.

In this chapter, I have concentrated on attempts to treat Alzheimer's by targeting amyloid beta. This focus makes sense both from the point of view that Alzheimer's has really *always* been a story of amyloid plaques (and tangles) and from the perspective that most of the headline drug trials to date have indeed involved antiplaque drugs. But this approach is somewhat misleading in that drugs with different targets are being developed all the time (the anti-Alzheimer's drug is the holy grail). Some have tau and tangles in their sights; others seek to inhibit or otherwise diminish the activity of the enzymes that improperly slice the amyloid precursor protein, APP, into toxic pieces. But so far, nothing remarkable has emerged from trials of these alternatives.

So it might be worth taking a step back to ask whether any of the approaches underway at the moment will generate an effective treatment in the end. Are we really looking at the complete picture of Alzheimer's? Everything I've described so far is based on plaques and tangles, their years of life history and the damage for which they are thought to be uniquely responsible. However, I mentioned earlier other mechanisms that show up at some point in the development of the disease—including inflammation, insulin resistance and mishandling of cholesterol. Are these simply signs of a brain that's breaking down, dragged into that state by plaques and tangles or are these other mechanisms implicated in causing the disease? It's not a cheering thought to think that there's still much work to be done to clarify the picture, but the history of science would warn us that this is likely.

Researchers are not just trying to stave off the impending health care tsunami; they're also seeking ways of making early diagnoses. As I was writing this chapter, an announcement was made about a panel of ten biomarkers, chemicals that become deranged by Alzheimer's, which could predict with 90 per cent accuracy the onset of Alzheimer's within three years.[7] Of course, such hopes have been raised, then dashed, before, so until this study is replicated on a larger scale with a more diverse subject group, it remains at best tentatively promising. And as long as there's no effective treatment, the prediction of "Alzheimer's within three years" seems arid. If it were twenty years in advance, some lifestyle improvements might help, but three years isn't long enough unless there's an effective treatment. Hence the rush.

Just to illustrate the likelihood that a lot of loose threads need to be tied up, I'll close this chapter with a description of a technique for alleviating the symptoms of Alzheimer's that comes at the problem from a completely different direction. It's called "deep brain stimulation."

This is a technique that has been used for at least twenty years to help people suffering from debilitating epileptic seizures or those with Parkinson's disease who have to endure uncontrollable movements of their limbs. In both cases, a slender wire electrode is implanted in targeted areas of the brain known to be involved in the problem (and which may be causing it). The wire is connected under the skin to an implant not unlike a heart pacemaker that generates electrical impulses. The effects can be striking: I've seen videos of Parkinson's patients sitting in a chair, arms flailing, but then when the juice is turned on, they become calm and get up and walk out of the room. Patients with epilepsy so severe that they were incapacitated have been able to return to normal life because of this treatment. But Alzheimer's?

Andres Lozano at the University of Toronto is at the centre of this research.[8] He and his team accidentally discovered that deep brain stimulation could enhance memory. They had implanted electrodes in a single patient, an obese man, who had failed after several attempts to control his weight. The thought was that stimulation of the hypothalamus, a structure intimately involved in appetite, might override whatever inappropriate signals it was producing.

The surprise came as soon as the electricity was turned on because the man began to report detailed memories of scenes from earlier in his life—scenes in which he had been an observer, not a participant, but otherwise consistent and peopled by those who had actually been present in those scenes. Somehow, the hypothalamus seemed to be connecting with the memory centres, especially the hippocampus, although this had not been anticipated. Lozano then set up a tiny six-person study to see what would happen if he stimulated either the hippocampus itself or neighbouring structures more intimately connected with it than the hypothalamus. All six people had Alzheimer's disease; all six were taking the anticholinesterase drugs described earlier in this chapter.

After a year of electrical stimulation, the results, while mixed, were intriguing. Some of the patients experienced the same kinds of vivid, reawakened memories—gardening, out on a lake "catching a large green and white fish"—that the single earlier subject had tapped into. The patient who experienced the most vivid of such memories also responded best from a symptomatic point of view. Two patients declined less after a year than would have been predicted. The other four didn't benefit much, so from the point of view of preserving cognitive function, this small study didn't provide strong evidence of effectiveness.

On the other hand, striking and long-lasting changes occurred in glucose metabolism, which is known to decline in Alzheimer's, especially in the so-called default circuits of the brain. These areas are active when the mind is "idle" (when we're daydreaming, for instance), and they're also the places where plaques prefer to congregate. Normally, in the progression of the disease, glucose metabolism in these areas steadily falls, but deep brain stimulation countered that drop and did so throughout the year of treatment.

Nobody, not even Andres Lozano, would suggest that deep brain stimulation is the next sure treatment for Alzheimer's. For one thing, it's hard to see how it would be possible to equip every Alzheimer's patient with the necessary gear. But these early findings (and more and bigger trials are now underway) underline the fact that the brain is both molecular and electrical. Yes, transmitter molecules conduct impulses from one neuron to the next, but the impulses within those neurons are electrical. It seems mysterious that stepping up the electricity can help while neurons are dying all around, but this discovery has made some begin to think of Alzheimer's as a "circuitopathy." Of course, as always, even hints of success raise questions like "Which part of the brain should be targeted?" In Parkinson's disease, where deep brain stimulation has been successful, the most damaged part of the brain is the substantia nigra, but the electrodes used in treatment target other, downstream areas that are not yet as grossly damaged. The same analysis would likely apply to Alzheimer's. Regardless, given where we are in the search for treatments at the moment, possibilities like this might turn out to be very important.

As I was *rewriting* this chapter, more evidence arrived, indicating that things are happening *extremely* quickly on this front.[9] Researchers at the University of Pennsylvania revealed that citalopram, a drug already available on the market as an antidepressant, both reduced levels of amyloid beta and stopped the growth of plaques in mice and, more important, reduced the levels of amyloid beta in cerebrospinal fluid in humans by close to 40 per cent. This drug, like Prozac, prevents the neurotransmitter serotonin from being removed from the synapse, thereby prolonging its activity—not unlike the first generation of anti-Alzheimer's drugs. Somehow (again reinforcing the idea that everything in the brain is connected to everything else), serotonin activity is connected to the continuing deposition of amyloid beta. A helpful discovery for certain—and the fact that the drug is already deemed safe and is on the market is a huge bonus. However, it's not clear yet whether reducing amyloid in this way in humans would translate into preservation of cognitive abilities. The next step is to test the drug in older humans with an eye to seeing whether actual plaques, not just amyloid beta, can be eliminated.

And finally, a study claiming the actual reversal of memory deficits in a small number of patients was published in the fall of 2014. The treatment comprised a daunting array of behavioural and dietary regimens, including taking supplements that in some cases totalled more than two dozen. So far, the study has too few patients to be celebrating, but the science is worth another look. [10]

CHAPTER FIFTEEN

Men, Women and Alzheimer's

The 2014 *Alzheimer's Disease Facts and Figures*, published by the U.S.-based Alzheimer's Association, includes a "Special Report on Women and Alzheimer's Disease."[1] Most of it details the incredible burden faced by caregivers, the majority of whom are indeed women. But the report also makes the point that there are differences between men and women who have the disease. These differences may be subtle, in that the general course of the disease seems distressingly similar in both genders, but sometimes slight differences like these open doors to new ideas.

Two-thirds of the Alzheimer's patients in North America are women. A striking statistic but one that doesn't necessarily mean women are more susceptible to the disease. Women live longer than men and therefore overpopulate the most vulnerable age group. A man who dies of a heart attack at seventy might have gone on to develop the disease (or in fact might even have been in the early stages but remained undiagnosed). In the 2014 report, the Alzheimer's Association is categorical: "There is no evidence that women are more likely than men to develop dementia at any given age." But that opinion is not unanimously shared: a minority of studies have presented data suggesting that women are indeed at greater risk. The resulting uncertainty is perhaps reflected in an earlier line in the same paragraph of the 2104 report: "The observation that more women than men have Alzheimer's disease and other dementias is *primarily* explained [italics mine] by the fact that women live longer, on average, than men, and older age is the greatest risk factor for Alzheimer's."[2]

While the studies purporting to show a difference in susceptibility between the sexes are agnostic about the cause, there is no doubt that differences exist, both chemical and physical, between men's and women's brains. So it would not be shocking to find that some of those differences might predispose one gender or the other to Alzheimer's. However, writing about the differences between women's and men's brains is fraught with challenges. Anyone who dares to extrapolate from gender differences in brain size or structure to differences in *behaviour* is inevitably going to come under attack. Human nature isn't so clear-cut, and attributing behaviour to innate differences in brain structure has an unfortunate history of being used to justify and maintain social inequities. At the same time, it would be very surprising if brain differences didn't make *some* contribution over and above culture and socialization.

The most comprehensive analysis to date of the physical differences between men's and women's brains found that overall, men's brains are bigger by anywhere from 8 to 13 per cent.[3] The amount of grey matter, white matter (the bundles of neurons sheathed in myelin), the size of the cerebral hemispheres and the total number of fluid-filled spaces within the brain all fall within that percentage range. It should be noted that it wouldn't be unusual to find a 10 per cent difference in brain size in a group of all men or women, as human brains vary widely in size. It's also true that brain size alone is not a predictor of intelligence. However, overall, there is a gender difference in brain size. What's more interesting is how that difference varies from one part of the brain to another.

Here, the pattern is anything but consistent. A complete list of the differences would be of interest only to neuroscientists, but the male brains in the studies compiled in this analysis (from ages zero to eighty) were either larger or had a higher density of cells in a total of sixteen parts of the brain, while the female brains had higher volumes or densities in fourteen. So what? Given that some of the areas that are different in men and women are prime targets for Alzheimer's—such as the hippocampus—and given that these areas eventually shrink with the loss of neurons, it's not inconceivable that size at the outset might be relevant. After all, the concept of brain reserve rests, in one interpretation anyway, on size.

The authors were careful to avoid the trap of linking these size differences to behavioural differences but did go so far as to argue that psychiatric disorders that are more prevalent in one sex or the other might be related to these findings. But they did not extend that claim to Alzheimer's.

However, another recent study strode confidently right into the size = behaviour trap and gained instant notoriety as a result.[4] This team used an MRI technique, called "diffusion tensor imaging," to make visible the main circuits of neurons in the brain, the connectors among brain areas. It's sometimes called "tractography." In a group of 949 youths between the ages of eight and twenty-two, the researchers found major gender differences: the strongest connections in males tended to stay within the hemispheres (that is, aligned back to front within the right and left), whereas females had stronger connections *between* the hemispheres. Only in the cerebellum, a part of the brain at the back densely populated with neurons, were these trends reversed.

So far so good, and actually really interesting. The researchers were able to parse their data to show that the male patterns were established solidly and early, but the connectivity in female brains developed more gradually from adolescence to adulthood. After presenting these data, the authors permitted themselves this definitive statement: "Taken together, these results reveal fundamental sex differences in the structural architecture of the human brain." Fair enough: it's a curious and interesting result.

Perhaps emboldened by the clarity of the data (and likely encouraged by the fact that the interhemispheric connectivity in female brains was reminiscent of previous findings), the researchers then expanded on the results: the connection from back to front in male brains "involves the linking of perception to action," they said, and that, in concert with the cerebellum, would create "an efficient system for coordinated action in males." Conversely, the cross-hemisphere connections in females would "facilitate integration of the analytical and sequential reasoning modes of the left hemisphere with the spatial, intuitive processing of information of the right hemisphere." Ah, so *that's* female intuition!

They then bolstered these observations with lab tests showing that the men were more adept in motor speed and spatial processing, while the females excelled in social cognition. So who could blame them for attributing those behavioural differences to the circuit maps they'd created?

Cordelia Fine could. She's a psychologist and writer at the University of Melbourne, and writing in the online journal *The Conversation*,[5] she attacked the study for "untested stereotyped speculation." She argued the circuitry differences might be simply a different wiring strategy for the larger brains of males, that the backup psychological tests actually revealed *at most* tiny differences between the sexes, and that the authors had ignored the possibility that the vast differences between social influences on girls and boys through childhood and adolescence might actually have shaped those brain differences. If the latter were true, these differences would not be hardwired. Fine labelled the conclusions "neurosexism."

Fortunately for me, I'm entitled to take a narrower view of this and share the report in light of the differences between males and females that might influence the development of Alzheimer's. For instance, it is known that tangles jump from one neuron to the nearest neighbour. Overall, the disease spreads along circuits, not popping up here and there randomly. But it's too early to say whether this finding has any direct relevance to Alzheimer's. So taken together, the most recent studies show that there are significant structural differences between men's and women's brains but nothing that would suggest a higher incidence of Alzheimer's.

However, even if the incidence of the disease proves, in the end, to be equivalent in both men and women, it would be very surprising to find that the *progression* of the disease was absolutely identical, given those brain differences. Good evidence has been found for differences in spread, and this could provide some crucial clues as to the nature of the disease. It could also suggest that when effective treatments become available, women should be treated differently than men. Both insights are important.

As the death of brain cells causes the amount of grey matter in the brain to shrink, the patterns of that loss differ between the sexes. It's definitely not a simple picture, but think of it as a map of the world where, as you watch, the borders of several countries begin to contract. Some of those countries border each other, and some do not (although, of course, in the brain, areas on the surface that appear to be unconnected might have direct neuronal contact under the surface). The bodies of water among them are the equivalent of

the spaces created as the surface of the brain shrinks: fissures widen
and hillocks pull away from each other. If this process was represented as a map of shrinking countries, those countries hardest hit
would be different in the male and female versions. This is also true
of the Alzheimer's brain, but it's not yet clear how such pattern differences affect the patient or indeed whether such differences could
lead to different approaches to treatment for the sexes.

Many such observations have come from ADNI, the
Alzheimer's Disease Neuroimaging Initiative, in which large
numbers of patients are imaged as they progress from mild cognitive impairment through to full-blown Alzheimer's. Researchers
can sift through the mass of data accumulated—including genetics
and changes in cognition and brain structure—and analyze the
data that interest them. One such study drawn from ADNI data
showed that in the early stages of Alzheimer's, women lose grey
matter (neurons) faster than men but that eventually such loss
accelerates in men to the point where they catch up.[6]

In line with these ADNI findings was a meta-analysis, a statistical compilation of fifteen studies showing that the impact of
Alzheimer's on women is more dramatic. The research team at the
University of Hertfordshire already knew that the Religious Orders
Study (see Chapter 9) had established that for a given amount of
pathology (plaques and tangles), women were cognitively much
worse off than men. But one study isn't enough. The fifteen studies
the team chose were the ones best suited to be grouped together and
analyzed statistically (it turned out that a relatively small number
qualified, something that surprised the researchers).[7]

Their analysis confirmed that women with Alzheimer's fare
less well on a battery of cognitive tests, the most striking of which
was a measure of verbal ability. (Prior to dementia, women consistently outperform men on such tests.) On the other hand, men

maintained their lifelong advantage on visuospatial tasks. Neither education level nor age was apparently responsible for the differences, but the authors did permit themselves some speculation. Although genetic testing hadn't been part of the studies analyzed, they wondered whether the well-known susceptibility of women to the notorious APOE4 gene might have played a role. Carrying one copy of the gene raises the risk for women much more than it does for men (although the impact of a double dose seems to be equivalent). However, the APOE4 gene is found in only roughly 20 per cent of the population, which limits its impact.

The authors also wondered about sex differences in cognitive reserve, but their most interesting thoughts revolved around the female hormone estrogen. The estrogen link to Alzheimer's is one of the most complex, tantalizing and potentially misleading issues in the entire story of men, women and Alzheimer's disease.

Interest in estrogen and its connection to Alzheimer's began decades ago. In women, the advent of menopause at about age fifty-one signals a dramatic drop in circulating estrogen. (Oddly enough, while women's life expectancy has been rising steadily since the mid-nineteenth century, the age of menopause has been remarkably stable, at 51.8 years.) While the first investigations were undertaken to see whether administering estrogen would enhance cognition in healthy postmenopausal women, one study in the 1980s took a crucial next step by pulling together some disparate conclusions from research that had already been carried out: that estrogen can influence behaviour in animals; that providing estrogen to female rats that had had their ovaries removed boosted the activity of the acetylcholine network in their brains; and that acetylcholine levels fall dramatically in humans who have Alzheimer's (the observation that spurred the development of the first anti-Alzheimer's drugs).[8] All that was enough to prompt the researchers to run a tiny clinical trial—administering estrogen for six weeks to seven women who had Alzheimer's. They noted improvements in attention, orientation and social interaction in three of the seven but cautioned that given the already well-known risks associated with hormone replacement therapy, it was much too soon to recommend this type of treatment for Alzheimer's.

The caution prompted by at best a modest balance of rewards to risks in 1980s studies gave way to greater enthusiasm and activity in the 1990s. Several studies in this decade showed that women who had taken estrogen for postmenopausal complaints seemed to have better resisted the onset of Alzheimer's than matched control individuals. It was still a mixed bag—some studies found no significant effect—but momentum was definitely being created. Here are some observations about the relationship between estrogen use and the onset of Alzheimer's (each of them from a different scientific paper of the late 1990s): "Oestrogen use in postmenopausal women may delay the onset and decrease the risk of Alzheimer's disease"; "[E]strogen replacement therapy may be useful for preventing or delaying the onset of Alzheimer disease . . ."; "Our finding offers additional support for a protective influence of estrogen in AD"; "There is also evidence that estrogen decreases the incidence of Alzheimer disease or retards its onset or both."[9]

The one caution-inducing factor was that none of these studies was truly definitive. A different kind of investigation was needed: a large, forward-looking (not backward-looking) trial that could reinforce the positive findings and relegate the negative to the back pages of science. And that's exactly what happened: the U.S.-based Women's Health Initiative was established in 1991 to examine the causes of illness and death in postmenopausal women.[10] At the outset, the WHI focused on cardiovascular disease, cancer and osteoporosis. But interest in dementia prompted the group to set up two different trials: the first, estrogen only for women who had had their uterus removed, and the second, a combination of estrogen and progestin for women with an intact uterus. (The estrogen-progestin combination was intended to prevent endometrial cancer.)

The results of the trials were a crushing blow to those who had hoped that estrogen might be a magic bullet targeting the huge problem of Alzheimer's disease in women. The estrogen-progestin therapy trial was halted early in 2002 because it induced a higher risk of heart attack, blood clots, stroke and especially breast cancer (something like a 26 per cent increased risk). The estrogen-only trial continued until 2004, but it was halted at that point too, having shown little benefit and some slightly increased risks, especially of stroke.

Most important from the Alzheimer's point of view was the demonstration of a *heightened* risk of Alzheimer's disease among women taking estrogen, alone or in combination. The results took scientists by surprise, but they were unequivocal. So was the conclusion voiced in the *Journal of the American Medical Association*: "Use of hormone therapy to prevent dementia or cognitive decline in women 65 years of age or older is not recommended."[11] The optimism of the 1990s vanished, virtually overnight.

However, as is always the case, even with definitive results, as these appeared to be, there can be qualifications. One that has come to be the focus of much attention is the fact that the average age of women in the study was sixty-three, and the average time since menopause twelve years, possibly too far removed from the onset of menopause for hormones to be effective. The researchers suggested that by that point, "irreversible neurodegeneration" might have set in—neurodegeneration in the sense of being unresponsive to estrogen.

This conclusion, although a decade old now, is not in dispute. Hormone replacement therapy is not useful for women in their sixties, for dementia or anything else. On the contrary, it could be dangerous.

So women *of that age* who either stopped ongoing hormone therapy or chose not to begin it likely benefited from the wide publication of the results. At that age, estrogen is not recommended. But that might not be true of younger women, who, frightened by the publicity surrounding the WHI results, made the same decision. Younger women were not part of the WHI study and shouldn't have assumed that the results, so clear for older women, were going to apply to them.

So what to make of all this? Before I outline an idea that could make everything clear, it's important to realize that one of the major factors in this sort of scientific uncertainty consists of the differences in the way studies are conducted. Ideally, researchers want to be able to assign treatment versus placebo groups randomly from square one, but that simply isn't always possible. Since the subjects in the WHI study ranged in age from sixty-five to seventy-nine, it would have been nearly impossible to be confident that the control group and the treatment group were equivalent.

Here's an example of how this uncertainty could arise: women who do the research and choose to start hormone replacement are generally better educated and healthier and already lead a healthier lifestyle. Some aspects of a less-than-healthy lifestyle, such as obesity and high blood pressure, are risks for Alzheimer's disease, meaning that the women with healthy lifestyles who chose to participate might already have had a lower risk for Alzheimer's. Then it would be hard to tell which was more important: healthy lifestyle or hormones.

Such biases are a significant problem when researchers try to reconcile the results of clinical trials: there are many ways of doing the experiments, and different methods can make the results of one study difficult to compare with others. That's the significance of the "randomized" in randomized studies: factors other than the treatment—factors you want to ignore (such as healthy lifestyle)—can be assigned equally to both treatment and placebo groups.

So first there were those studies in the 1990s showing that estrogen was effective in preventing or delaying Alzheimer's. Then the WHI study demonstrated exactly the reverse—at least for women in their sixties. It would have been easy to assume (and indeed thousands did) that the WHI results applied to *all* women who'd taken estrogen regardless of when they started estrogen therapy or how long they maintained it. But even well-designed studies like this deserve scrutiny. For instance, the higher risk of Alzheimer's for women over sixty taking estrogen is a relative risk: more cases of Alzheimer's relative to the background rate (that is, the numbers of Alzheimer's cases you'd expect, all things being equal). But the background rate is substantially lower in younger women, and if the higher estrogen-linked risk is applied to them, it would only mean something like one additional Alzheimer's case among a thousand women taking hormones for five years. But there were reasons beyond that to focus on the question of whether the WHI study was relevant for younger women.

First, aside from the studies of postmenopausal women, fundamental hormone biology and animal experiments in the 1980s and 1990s suggested that estrogen should be helpful. For one thing, the brain is tuned to estrogen, bearing receptors for the molecule both on the outside cell membrane and the membrane surrounding the nucleus of the neuron. As a result, estrogen can gain access to the DNA inside the nucleus and influence the expression of genes sequestered there. Estrogen is also known to be both protective of neurons (enhancing recovery from traumatic brain injury) and supportive of neuron growth.

Animal studies also suggested that the *timing* of estrogen deployment was important. For instance, estrogen regulates the growth of spines on neurons in the hippocampus. If a female rat's ovaries are removed, estrogen plummets and the spines wither. If estrogen is started four days later, the spines recover, but if the treatment is delayed for another eight days, they don't. That's the physical side: when rats without ovaries are challenged by memory tasks, they do better when given estrogen three months later, but not if it's administered twelve months later.[12]

Out of this welter of conflicting results has emerged a theory that might explain them. It's called the "critical period" hypothesis, and it argues that there might be an ideal time, at the beginning of menopause, during which the administration of estrogen might help forestall dementia, but that this period is limited and hormone replacement should be ended after four or five years.

The critical period hypothesis has been on people's minds for a decade now and has turned out to be a robust explanation of the differences between the Women's Health Initiative's negative findings and the several earlier, positive ones. Subsequent studies have shown that women who had begun hormone therapy for various reasons at relatively young ages performed better years later on memory tests. This result did not prove that hormone therapy initiated at an early age would prevent Alzheimer's, but it is suggestive of it. One of the most convincing studies (convincing because it was randomized) involved more than 260 women receiving estrogen to stave off osteoporosis. The women started at menopause, stopped two to three years later and were then tested anywhere from five to fifteen years after that.[13] They were at much lower risk of cognitive impairment than were the placebo-taking participants. But this issue is far from settled: a study announced in 2013 found no cognitive benefit among women in their mid-sixties who had taken estrogen in their fifties and then stopped.

The existence of a critical period at a younger age during which estrogen might work would tally nicely with the prevalent idea that in Alzheimer's, the disease process is well underway before any signs are actually visible. Of course, with the history of conflicting reports having haunted this area of research, more, and more definitive, studies will be needed. But it wouldn't be shocking if in some future large study, researchers recruited women just entering menopause, administered estrogen for three or four years, then waited a decade or two and revealed a beneficial effect. "Beneficial" being at the very least a delay in the onset of Alzheimer's if not something more dramatic.

CHAPTER SIXTEEN

Was It Really the Aluminum?

I f you lived through it, you'll remember it. Aluminum was responsible for Alzheimer's. We learned that we shouldn't steam acidic foods like rhubarb in an aluminum pot; that aluminum in antiperspirants would leach into the skin, then the blood, cross the blood-brain barrier and settle in with some welcoming neurons; that cigarette smoking imported aluminum into the lungs. Even that most innocent of beverages, tea, was viewed with suspicion. The concerns started in the 1970s and carried through at least to the 1990s, but if you cruise the Internet today, you'll see that many still shy away from aluminum because it "leads" to Alzheimer's.

When I started this chapter, my goal was straightforward: I was going to say that aluminum as a contributor to Alzheimer's is barely on the scientific radar, that reports of the connection between aluminum and Alzheimer's are few and far between and that the scare seems to have evaporated. So what happened? How did science cast doubts on the aluminum threat?

That was my goal. It didn't take me long to abandon it because it's not all that clear *what* happened. One thing is sure, though: a lot of science addressed the aluminum question. The investigations started relatively innocently, driven by scientific curiosity. Then they gained momentum as the circle of investigators widened, and at some point, with confirming evidence apparently not as robust or consistent as it could have been, proponents and opponents sorted themselves out and seem to have fought to a draw. But a draw, a mix of evidence pointing in different directions, isn't enough to mobilize scientists. Here's how it went down.

"The origin of this study is rather accidental."[1] This statement was not only a most unusual, charming and self-effacing way of beginning a scientific paper; it also had the distinction of sparking the long, dramatic, twisting-and-turning scientific saga surrounding the possibility that aluminum was connected with Alzheimer's. At one time, the paper it introduced appeared to be part of a masterful exposé of one of the major causes of Alzheimer's disease, but this is not so much the case today.

It was 1965, a time when the disease itself had not taken on the dread aspect it has today (though plaques and tangles and their involvement in Alzheimer's disease were already well known). One of the authors of the report, Igor Klatzo, had already distinguished himself as a researcher by connecting the dots between the mysterious disease in New Guinea called "kuru" and the human dementia, Creutzfeldt-Jakob disease, a piece of scientific detective work that I described in *Fatal Flaws*. The second author, Henry Wisniewski, went on to become a central figure in Alzheimer's research until his death in 1999.

But this scientific paper wasn't actually *about* Alzheimer's disease. The Klatzo research group was investigating the strange case of epilepsy in rabbits caused by inoculations of an aluminum salt called "alum phosphate." A by-product of the onset of seizures in the rabbits was the appearance in their brains of twisted strands of protein, recognized immediately by the researchers as being just like the tangles seen in the brains of Alzheimer's patients, at least superficially.

The study was a fairly straightforward examination of the events that transpired between the injection of the aluminum into the brain and the appearance of the filaments. The next report in the same journal pushed the investigation further but produced an "on the one hand but on the other" sort of conclusion. That is, yes, these so-called neurofilaments did resemble the tangles in Alzheimer's disease in several ways, but in the electron microscope, there were obvious differences in size and shape, leaving the investigators calling for more research to see whether the two were definitely related or not. They also pointed out, importantly, that the rabbit tangles were always bound up with aluminum and so were probably caused by it, but there was no known association between aluminum and Alzheimer's disease.

Even given the apparent differences between the two kinds of tangles, it was too tantalizing a coincidence to ignore. As second author Wisniewski put it, "After this publication, the whole world was talking about aluminum as a possible cause of Alzheimer disease . . ."[2] and added that it took years for him and others to show the great differences between the tangles in the rabbits and the tangles in humans with Alzheimer's. Before that happened, these studies triggered an entire branch of science, and some real public fear, about the risks of aluminum. Those fears have not entirely disappeared.

Among those who took note of these intriguing initial results was a Canadian scientist, Donald Crapper McLachlan, who was to become a central figure in the aluminum-Alzheimer's chronicles. Inspired to work on the issue by the Alzheimer's-like qualities of the tangles in the brains of epileptic rabbits, Crapper McLachlan, realizing, of course, that this resemblance wasn't enough to link the two, broadened his view.

He started by checking out the aluminum levels in autopsied brains, both with and without Alzheimer's disease.[3] It must have been a slightly tedious business, extracting tiny pieces from different areas of the brains and measuring the amounts of aluminum in each. Crapper McLachlan noted that extreme care had to be taken to ensure that no aluminum from the surroundings could make its way into the samples, an important issue, given that this is the most common element on earth. Many kinds of contamination had to be prevented: from soap, paper towels, tap water, room dust and even—a sign of the times—cigarette ash.

The results suggested that the aluminum connection was worth pursuing. The average amounts of aluminum in Alzheimer's brain tissue were double those in non-Alzheimer's brains, and the accumulations were greatest in areas of the brain with the most tangles. But there was no correlation with the distribution of plaques in those brains. Adding those observations to evidence that animals with similar brain concentrations of aluminum exhibited behavioural deficits was enough to provoke curiosity and create momentum among researchers, especially since the first report to this effect appeared in the much-read journal *Science*.[4]

Curiosity and momentum maybe, but no alarm bells yet. Animals injected with aluminum accumulated strange agglomerations in their brains that looked something like—but not completely like—the tangles of Alzheimer's, and a small sample of human brains with the disease seemed to have higher levels of aluminum than normal, tending toward the places where there were the most tangles. Not a smoking gun yet—just a small set of intriguing, but unconnected, observations.

But then things started to heat up. Americans Daniel Perl and Arthur Brody showed that in addition to more aluminum residing in areas of the brain dense with tangles, aluminum appeared to be clustered close to the nuclei of neurons inside of which tangles were developing. This was precision that Crapper McLachlan had been unable to attain with his methods.[5] The sample size was tiny (three brains with Alzheimer's and three without), but the results were startling: 90 per cent of neurons in the hippocampus with tangles contained aluminum, whereas adjacent neurons free of tangles had almost none.

Perl and Brody's paper was a model of restraint, but what followed wasn't. A psychiatry resident at Yale named Steven Levick published a letter in the *New England Journal of Medicine* suggesting that kitchen cookware might be slowly poisoning large numbers of Americans—all on the basis of Perl's work and on the observation that the cheap aluminum pots and pans Levick had purchased as a student had begun to show signs of pitting and wear and tear after only a couple of years' use.[6] The idea of slow poisoning with Alzheimer's waiting in the wings attracted immediate attention in media around the world. Note that Levick was making a suggestion only: he had no data other than Perl and Brody's to back up his argument. Nevertheless, the aluminum controversy was soon underway.

But while the public was reacting pretty much the way you'd expect—expressing everything from panic to complete lack of interest—the world of science, in its own moderate way, was demonstrating its own interest by embarking on more experiments. In the years from 1986 to 1991, the following claims were published in rapid succession.

An English lab reported not only that aluminum was associated with tangles in Alzheimer's disease brains but also that it could be found in the cores of amyloid plaques. Of course, aluminum's involvement with plaques *and* tangles put the story on a firmer footing. This study attracted several letters to the British medical journal the *Lancet*,[7] one of which pointed out that it looked as if the aluminum likely entered the plaques late in their history, suggesting it was irrelevant to their development; the others tilted toward believing this was another piece of evidence implicating aluminum. And so an on-again, off-again story was generated that lasted for the next several years.

The *Lancet* was just warming up. In January 1989, it published a study of eighty-eight English counties showing that the rates of Alzheimer's disease where the aluminum concentration in the drinking water was greater than 0.11 milligrams per litre were 50 per cent higher than in places where aluminum concentrations were a mere 0.01 milligrams per litre. The research targeted drinking water because the aluminum in it (used to clarify the water) was thought to be absorbed by the body more efficiently than aluminum from other sources. Had I lived in one of those places, I would have been troubled by this report, but as always, there was caution: "The results of the present survey provide evidence of a causal relation between aluminium and Alzheimer's disease. However, care is needed in interpretation. . . ."[8] It's the usual scientific tension: you can't appear to have missed out on the potential impact of the research, but at the same time, you can't let your peers think you've overstated the case. Yet months later, a study in the southwest of France produced roughly the same results.

At about the same time, a group in Washington state reported it had found a weak relationship between long-time use of aluminum-containing antiperspirants and the risk of Alzheimer's but virtually turned around and dismissed its own findings because the data-gathering technique of questioning surrogates (often necessary when Alzheimer's patients are the subjects) had produced too much uncertain data.[9]

In 1991 Donald Crapper McLachlan, perhaps moved by the breadth and amount of new data, made a straightforward statement in the *Canadian Medical Association Journal*: "Would decreased aluminum ingestion reduce the incidence of Alzheimer's disease?"[10] He and several colleagues argued that reducing personal intake of aluminum would be a prudent public health measure, as according to their hypothesis, it would reduce the incidence of Alzheimer's. By now, the original discovery that aluminum injections caused epilepsy and tangle-like deposits in rabbits was nearly twenty years old. Crapper McLachlan added the evidence that aluminum caused cognitive deficits in animals, that it collected in those places in Alzheimer's brains where the most damage was found, that drinking water seemed to elevate risk and that a metal-binding drug, desferrioxamine, appeared to have preserved cognition in a group of patients over a period of two years, presumably by grabbing onto aluminum.

Here's what Crapper McLachlan concluded: "Four independent lines of evidence support the conclusion that aluminum is an important risk factor in AD [Alzheimer's],"[11] but he went on to admit that none was absolutely rock-solid. The term "risk factor," for instance, nibbles at the heart of the controversy: Was aluminum the cause of the disease or did it simply collect once the disease was established? The difference is crucial, and even by this time, twenty years after the original data, that question hadn't been answered.

It was at this point (in about the early 1990s) that any trend toward establishing aluminum as a definite risk factor for Alzheimer's slowed and then began to be plagued by harsh criticism and weakened by contrary reports. It would never regain its momentum. And as sometimes happens, after years of debate and research, it's one study, one published report, that turns things around. That's the impression you get from the reception to a short letter from Oxford researchers to the journal *Nature* in late 1992.[12]

These scientists used new techniques to throw cold water on the idea that aluminum was present in Alzheimer's plaques; their nuclear microscopy failed to find any but did reveal aluminum in the background, prompting them to suggest that no matter how careful lab workers are, there is always aluminum contamination: in instruments, stains used to highlight cellular features, everywhere. In their minds, the aluminum previously found in Alzheimer's plaques was simply that: contamination.

That suggestion couldn't have gone done well with the scientists who were, in effect, being told they'd been sloppy. But it had a much wider impact. Although the *New York Times* came out with the headline "New Alzheimer's Study Questions Link to Metal,"[13] Daniel Perl made what should have been an obvious point by reminding readers that he'd already provided solid evidence that aluminum was found around *tangles*. The fact that it might not be in plaques didn't change that and could be seen as irrelevant.[14]

Tempers appeared to be a little frayed at this point, as was demonstrated by a delightful little spat in the pages of the *Canadian Medical Association Journal* in 1994. It all began with a story by freelance writer Marvin Ross about the Alzheimer's Disease International Conference held in Toronto.[15] The story was generally positive about Donald Crapper McLachlan's work with aluminum, but the response was not. Dr. David Munoz, then at Western University in London, Ontario, blasted the journal for publishing a "biased, distorted account," claimed that "no scientific group other than Crapper McLachlan and his colleagues has advocated reduced exposure to aluminum" and questioned why the journal would publish an aluminum/Alzheimer's story when his search of the medical literature revealed only 15 articles on the connection between them out of a total of 3,803 on Alzheimer's itself.[16] Munoz described it as science that had been "left behind."

A flurry of responses, all supporting Crapper McLachlan, followed in that same issue of the journal, including one of his own, boldly claiming that reducing aluminum in Ontario's drinking water could eliminate tens of thousands of cases of Alzheimer's.[17] Munoz, apparently sensing that he had been piled on, shot back at the journal for "giving the impression of massive support for his [Crapper McLachlan's] views" and added that the *Canadian Medical Association Journal* "stands apart from other scientific journals in its support of fringe theories."

Gems like these reassure me that occasionally the material in scientific journals rises above a weary conclusion drawn by Francis Crick (of DNA fame) that "[t]here is no form of prose more difficult to understand and more tedious to read than the average scientific paper."[18] But aside from the entertainment value, what did that exchange say about the state of the aluminum hypothesis?

That was 1994, and while it's persuasive to read Munoz' stats that only 15 articles on Alzheimer's out of 3,803 actually dealt with aluminum, numbers alone don't necessarily represent the state of the research accurately. After all, in the middle of the twentieth century, there weren't very many papers being written on Alzheimer's disease itself. But Munoz wasn't alone in his skepticism (although few would have gone so far as to label the work of Crapper McLachlan and his colleagues as "fringe" science). It was still unclear exactly what role—if any—aluminum played.

Yes, there were data implicating drinking water and links between tangles in the brain and the metal, but several studies failed to support those connections. In fact, the team that reported the first significant connection between Alzheimer's and drinking water in England in 1989 reversed its stand in 1997, claiming that most previous studies implicating drinking water (including its own) were incomplete. This was partly because the studies hadn't taken into account varying levels of aluminum in water supplies over time and partly because silica, an incredibly common substance, which would interfere with aluminum uptake in the body, wasn't taken into account. So the team measured both factors and found no risk from aluminum, even with amounts that in previous studies had been associated with a greater risk of Alzheimer's. Seemed like no risk. But again, that ambivalence: "We cannot rule out an important association of Alzheimer's disease with aluminum in drinking water at very high concentrations, but our findings indicate that any relation at levels below 0.2 milligrams per litre is weak."[19]

Even the drinking water link seemed odd in that people who consumed certain antacids swallowed thousands of times more aluminum than they did in their drinking water, yet no association of Alzheimer's with antacids had ever been found. There were questions about how easily aluminum could get into the brain (especially from the stomach), why there was so much variation in results from study to study and where exactly aluminum fit into the disease process of Alzheimer's. By this point (in the mid-1990s), the amyloid cascade hypothesis had taken hold, and given that it argued that plaques come first, followed by tangles, how could aluminum be causative when it might be associated with tangles but not with plaques?

And so the debate went on, neither side really gaining anything more than a temporary advantage. In "Controversies in Neurology," a little cut-and-thrust piece in the journal *Archives of Neurology*, the aforementioned Dr. David Munoz returned to the fray, apparently having lost none of his disdain for the aluminum hypothesis.[20] His essay is decorated with phrases like "Mainstream science has long ago left behind the aluminum hypothesis, which is generally considered a fringe theory" and my favourite, "The *Lancet* once published an article titled 'Aluminum Hypothesis Lives.' So does Elvis, according to some." *Bon mots* aside, Munoz is convincing in his dissection of the case for aluminum. But of course, "Controversies in Neurology" presents a counterpoint, this one by William Forbes and Gerry Hill. It's a muted comeback, "a qualified yes,"[21] but they urge patience, arguing that the aluminum story is too complex for a rush to judgment. Its variables include the need for detailed and accurate diagnoses of Alzheimer's, the chemistry and impacts on the brain of the various forms of aluminum and, in the end, the statistics required to evaluate all of this evidence and analyze widely disparate groups of patients. The authors present a long list of items to be resolved but conclude that "aluminum in water is one possible risk factor for developing AD."

Rhetorically, Munoz might have won the day, but the aluminum story did not die, even though scientific interest in it was, in the absence of dramatic findings, waning. It continued to live at least partly because the web of evidence, even if each piece wasn't particularly robust, seemed to signal that something was afoot. Then in 2006 a strange case was reported from England.[22] It concerned a woman who lived in Cornwall, in the area where a massive amount of aluminum sulphate had spilled into the drinking water—twenty tonnes of it. That had been back in 1988. In 2003, when the woman was fifty-eight, she was experiencing symptoms of dementia: word finding, naming objects and simple calculations had become extremely difficult. She steadily deteriorated and, sadly, died in 2004. At autopsy, multiple amyloid beta deposits were found in the blood vessels in her brain but few amyloid plaques. However, the most astonishing thing was that she had levels of aluminum in her brain twenty times higher than average.

This case is simply the story of aluminum and Alzheimer's—in miniature. There is dementia and there is definitely excess aluminum, but few of the diagnostic features of true Alzheimer's disease (like plaques and tangles) are visible. The woman from Cornwall also had two copies of the APOE4 gene, meaning that she was at risk for Alzheimer's anyway. There's also no real way of comparing this amount of aluminum in the brain with the levels one would normally find. As many as twenty thousand people were exposed to the aluminum released in the spill—and perhaps more cases will surface. It's worth noting that the investigation that followed the spill was not initiated immediately, and some have argued that a mix of media frenzy, conspiracy theories and litigation made it impossible to find local individuals for study who had no bias. It's also been argued that many of the reported symptoms, like memory loss and fatigue, would have been found before the spill as well.[23] Of course, an immediate follow-up says nothing about delayed impacts. This one case could signal something, but until there are others like it, it's hard to know what to do with a single-person observation.

So where are we now, well into the twenty-first century? The investigation of aluminum and its connection to Alzheimer's has definitely lost whatever momentum it had twenty years ago. Too many conflicting results, hints that remained unresolved, more promising paths on which to base a scientific career: all of these have had their impact. Papers are still being published. In one, the author, Lucija Tomljenovic, points out that the use of aluminum in drinking water to reduce turbidity and organic matter began in the late 1880s and that the first Alzheimer's case was roughly twenty years later. An association! She points out that a

1926 article described Alzheimer's as being rare, something that might make sense if aluminum ingestion needed to be long-term and chronic before a rise in Alzheimer's would be seen.[24] But as I pointed out in Chapter 3, attempts to estimate the actual number of cases of Alzheimer's at any point in the twentieth century by counting medical journal references can never be accurate: dementia wasn't considered to be a medical issue, and even if it were, it was attributed largely to circulatory issues. (Tomljenovic finds space in her article to take swipes at fluoridation of water and vaccination as well.)[25]

Two recent papers out of Japan argue that yes indeed, after all these years, there *is* evidence of aluminum in plaques and that there are ways one could accommodate aluminum in the famed amyloid cascade hypothesis.[26] (I mention Japan because back in the 1990s, this was a Canadian story. More was written on the possible aluminum-Alzheimer's connection by Canadian scientists than by researchers from any other country.)

Is aluminum a risk? I'm left with a vague sense of discomfort, not being able to either embrace the hypothesis or to dismiss it completely. If I had to bet, I'd say that in the end, even if aluminum is shown to play a role in the development of Alzheimer's, that role will not be significant compared to many others that I've mentioned, including education, brain size, conscientiousness, blood pressure, atherosclerosis, diabetes, obesity, physical fitness, mentally stimulating work—and likely many more.

CHAPTER SEVENTEEN

The Many Faces of Dementia

You don't have to venture very far into Google to find the human brain described as "the most complex object in the known universe" or something like that. But for all we know, you wouldn't have to go any further than the nearest whale to poke holes in this claim. Anyway, I'm not here to quibble but to add the follow-up to that description: "It shouldn't be surprising that there are myriad ways it can go wrong."

Throughout this book, I've concentrated on Alzheimer's disease as the most prominent and troubling form of dementia. It certainly is the most common: estimates of its prevalence among all dementias hover between 65 and 75 per cent. Its sheer numbers make it the most worrying from a health care perspective, but the other 25 per cent or so of dementias are not insignificant, not only for the individuals and the families involved, but also for the clues to dementia in general that might be gleaned from them.

Take Creutzfeldt-Jakob disease, for instance. Aside from some genetic forms, it is unpredictable and certainly much rarer than Alzheimer's, with an incidence of one case per million people per year. It was first diagnosed at roughly the same time that Alzheimer announced his disease to the world but really didn't reach public consciousness until the mid-1990s, when a new version, *variant* CJD, appeared in England. To date, there have been more than two hundred cases of this particularly harrowing dementia, mostly in the U.K., caused by ingesting the prion that caused the mad cow disease epidemic in the 1980s.

The good news is that variant CJD has practically disappeared, the result of interrupting the chain of infection that fuelled the mad cow epidemic to begin with. But the parent disease CJD continues on, its rate largely unchanged over the last several decades, with no evidence whatsoever of the involvement of an infectious agent. Creutzfeldt-Jakob disease is, perhaps mercifully, much quicker than Alzheimer's, as most patients die a few months after diagnosis, but it has some similarities as well that extend beyond the mere fact that both involve the derangement of cognitive abilities.

One study concluded that as many as 13 per cent of cases diagnosed as Alzheimer's turned out to be CJD (although that number seems improbably high).[1] More important, there are similarities in the way the two diseases spread in the brain. Prion diseases like CJD build momentum as aberrantly folded versions of a made-in-the-brain protein conscript normal versions by forcing them into the misfolded state as well. Eventually, these aberrations disable the brain. That notion of misfolding, recruitment and spread has its counterpart in Alzheimer's disease, where lab experiments have convincingly shown that inoculating amyloid-rich material from, say, a deceased Alzheimer's patient into the brain of a genetically engineered mouse can induce the multiplication and spread of plaques in that mouse.

The process is called "pathogenic protein seeding." It's important to note that the much shorter precursors of full-blown plaques, not the plaques themselves, usually trigger this activity. And a very similar process seems to drive the formation of tangles in Alzheimer's disease.

An all-important caveat about this research, though: whatever the resemblances at the cellular/molecular level between Alzheimer's and CJD (and other prion diseases), there is a crucial difference. The prion diseases are infectious, whereas as far as we know, Alzheimer's is not. There was, however, a flurry of interest and fear about two years ago when a suggestion was floated from the University of Texas that researchers there had managed to transmit plaques from mouse to mouse by blood transfusion. I was in a room at the time with a group of prion researchers, and this announcement caused a great deal of consternation.

The announcement, apparently premature given that it never appeared in a peer-reviewed journal, did at least lead indirectly to a meeting at the Roslin Institute in Edinburgh in November 2012, where experts approached the issue from several different directions, trying to find any scrap of data that might suggest that Alzheimer's is infectious. They couldn't uncover any such evidence. Nor did a study published shortly afterward by a team at the University of Pennsylvania. The subjects were what might seem to be an unusual group: people who had received human growth hormone (to compensate for a lack of their own), processed from the pituitary glands of the recently deceased.[2]

There's reason to be concerned about the possibility of infection because in the 1970s and 1980s, more than two hundred people died from Creutzfeldt-Jakob disease transmitted by infectious prions in the pituitary. (U.K. prion researcher Alan Dickinson tells the story of how one night he had a terrible thought about human growth hormone: if even one of those pituitaries was from a patient who had died of CJD, many young children would be at risk for the disease.)[3] The math was done, and one estimate suggested that one pituitary in a thousand might have been infected. Hundreds of thousands of glands were processed, meaning that *hundreds* of infected glands could have entered the system. No one can confirm that number, but some pituitaries were infected because those two hundred people died as a result of being injected with tainted hormone.

So the U. Penn. team tackled this issue from the front end *and* the back end. They examined pituitary glands from autopsies to get an idea of how many plaques, tangles and other signs of Alzheimer's pathology could be found. That was the front end. They also went to the medical records of the thousands of Americans who had received human growth hormone from autopsied pituitaries and scanned those records looking for suspicious deaths (not just Alzheimer's but also Parkinson's and Lou Gehrig's disease). About six thousand patient records in all; about eight hundred deaths. That was the back end.

There was no mention of either Alzheimer's or Parkinson's on any of the death certificates, but there had apparently been two cases of Lou Gehrig's disease (also called "amyotrophic lateral sclerosis" or "ALS"). Neither case had been as thoroughly examined as the researchers would have liked, but it didn't matter because two cases didn't reach significance.

They then uncovered a hitherto unnoticed case of a young man, also with ALS. With his case added to the total, the picture changed: three cases was more of an alarm bell than two. Not a piercing alarm, but a number that attracted attention. While the concern was about ALS, not Alzheimer's, evidence of infectivity of one would necessarily raise fears of infectivity of the other. This surprise third case was the attention grabber. He had died of ALS at eighteen, nearly twelve years after he received growth hormone. That might seem to suggest a direct correlation. But then some of his spinal cord tissue was inoculated directly into the brain of a capuchin monkey, which showed no ill effects and lived another eleven years. Not proof of lack of infectivity, but a strong suggestion.

But what about Alzheimer's itself? The researchers cautioned that there were still many unknowns. Having found no apparent cases of the disease did not mean that none would eventually appear. The incubation period, the time between being infected and actually showing symptoms, is the issue. In the case of Creutzfeldt-Jakob disease transmitted via growth hormone, the incubation period ranged from five to forty-two years! So in this recent study, no one can really predict whether some cases might still be brewing. The researchers also admitted that they didn't have the technology to analyze autopsy tissue to look for signs of the earliest phase of Alzheimer's. For both these reasons, the Alzheimer's incidence, which appeared to be zero, may actually have been higher.

A perfect study of the potential infectiousness of Alzheimer's would require decades of follow-up with people, some of whom had had close contact with an Alzheimer's patient and some not. Care would also have to be taken to ensure that their other risks were more or less equivalent. That would involve statistically managing all factors, such as genetic and environmental influences, brain reserve and education—anything that could mess up the results. These days, it wouldn't even be easy to find a control group, a set of people who have had *no* contact with the disease. Such a study will likely never be done, but based on the evidence that does exist, I'm definitely not worried about "catching" Alzheimer's.

That similarity between CJD and Alzheimer's, at least at the molecular level, exists among other dementias and neurodegenerative diseases as well, including frontotemporal dementia,

dementia with Lewy bodies, even Parkinson's disease and ALS. Vascular dementia, which I've already discussed, doesn't operate on the same principle (although it does interact with Alzheimer's), and there's also the somewhat less well studied chronic traumatic encephalopathy, formerly known as *dementia pugilistica*, which may involve cell-to-cell transport of misfolded tau protein.

These dementias all cause cognitive decline, associated with many other symptoms, and it's all but inevitable that research will uncover similarities that aren't yet apparent. Whether any of these might accelerate the race for an Alzheimer's treatment is anybody's guess at the moment. But obviously, for the sake of patients with any dementia, the faster research moves, the better.

How widely should the net be cast? The science is challenging because you need to take risks and look at things that might not appear to be hopeful but confine yourself to the practicable at the same time. And then, every once in a while, a network of stories, all very different, suddenly connect. There's one condition that illustrates this perhaps better than any other. It also makes the point that a dementia can have multiple influences that defy explanation.

This story is set on the island of Guam in the South Pacific. Like so many other places, it has seen a succession of conquerors over the centuries, but now it's an American territory. The total population of Guam is somewhere between 180,000 and 190,000, of whom about 70,000 are the Native people, the Chamorros. The Chamorros attracted the interest of medical experts immediately after World War II when it was discovered that they were plagued by a pair of unusual neurodegenerative conditions: one that combined dementia and Parkinson's disease and the other a form of ALS. The rates of ALS among them were one hundred times greater than elsewhere in the world.

These conditions seemed at first to be unique to Guam, though tiny pockets of what appeared to be similar syndromes were subsequently discovered in Japan and the western end of the island of New Guinea. The unique qualities of the Guam versions made them worthy of interest to begin with, but the realization that both diseases were in rapid decline only piqued interest further. Why were the conditions disappearing? Where had they come from? Why were the rates of disease dramatically different when two towns less than ten kilometres apart were compared? Add to that the observation that a completely different, special Guamian version of Alzheimer's, with tangles but no plaques, was on the rise, and the spotlight was fully on Guam.

The decline was intriguing; diseases do wax and wane, but this was a *pair* of diseases, ALS and Parkinson's. ALS was first recorded right after the war and held steady until about 1960. Parkinson's didn't appear until the mid-fifties but climbed rapidly and soon matched ALS. Then they both started dropping. At a time when the general population was increasing dramatically, both ALS and Parkinson's were dropping like stones in a pond. In only a decade, fewer than half as many cases of both were appearing every year. Today, this version of ALS is rare on Guam: not a single person born after 1951 has come down with the disease.[4]

When a decline is this rapid, it won't be a genetic fault but likely something environmental. And guess which environmental factor was first suggested in this case? Aluminum. The same Daniel Perl who was active in the aluminum and Alzheimer's controversy found abnormally high levels of aluminum in neurons in the brains of Guamians that contained tangles, regardless of whether the patient had ALS or Parkinson's.[5] Again, an intriguing finding, but not unlike the aluminum hypothesis itself, this side branch withered away.

Then the story took the first of several steps off the beaten path. In the early 1960s, cultural anthropologist Marjorie Whiting suggested that an important ingredient in the Chamorros' diet, the seeds of the false sago palm, might be responsible.[6] This plant is one of the cycads, palm-like plants that were so dominant in the past that a true Jurassic Park would have been carpeted by them. The seeds, which are about the size of a ping-pong ball, are toxic, and the Chamorros, well aware of this, put them through several rounds of soaking and rinsing in water before grinding them into flour. Nonetheless, the flour retained some toxicity, enough to suggest to Whiting and others that this strange combination of neuro-degeneration might be the result of poisoning. The decline in incidence fit nicely with the theory: during World War II and Japanese occupation, the Chamorros had little or no access to rice and had to depend on cycad flour. After the war, consumption of cycads dropped steadily. Allowing for the years it takes to establish disease, the timing fit well with the decline of the diseases after 1960.

But the fact that the Chamorros processed the seeds specifically to reduce the level of toxins made the connection suspect. It seemed impossible that the people could consume enough seeds to poison themselves. Problematic too were these two facts: animals fed relatively huge amounts of cycad didn't develop any signs of neurological disease, and there were no cycads eaten in the two other pockets of the diseases, Japan and Indonesian New Guinea.

A second off-beat step was required, and living legend Oliver Sacks stepped in to provide it. The author/neurologist teamed up with Paul Cox, an ethnobotanist, and argued that there was one way of concentrating the cycad toxin. "We suggest," they said, "that the Chamorro population of Guam ingested large quantities of cycad toxins indirectly by eating flying foxes."[7] Flying foxes. Yes, that's what they said. Bats.

Their thesis relied on the idea of biomagnification, the same principle that makes it unwise to eat certain ocean fish because they concentrate long-lasting chemicals from lower in the food chain, especially in their fat. Chemicals flow through millions of algae, thousands of tiny fish, hundreds of medium fish, a few large ones and may end up in a single whale. Flying foxes are the cuter version of the two main kinds of bats. They're bigger than the North American versions, with faces that do indeed look like foxes', and at one time, they flourished on Guam. Again, unlike their North American cousins, they eat fruit and nectar (rather than insects). The Guam versions love cycad seeds, from which they suck the juice and spit out the pulp. Before flying foxes were hunted nearly to extinction (partly because of the postwar influx of guns), flocks of them would forage among the cycad trees on the island. Flying foxes are also known to accumulate toxic chemicals in their body fat.

The next piece of the puzzle: Chamorro people dined on flying foxes. They boiled them in coconut cream and ate them whole, fur and all. Two of the villages that were known to indulge in this practice most enthusiastically were also two of the villages with the most neurodegenerative disease.

So the first explanation for the decline of ALS and Parkinson's was that once rice became available, dietary habits changed; people just stopped eating cycads. Sacks argued that no, they stopped eating cycad products because bats became so rare (one species was extinguished completely). No bats, no bioconcentrated toxin, no disease.

Sacks's co-author Paul Cox went further by measuring levels of specific toxins both in cycad seeds and in three skins of flying foxes from Guam preserved in the Museum of Vertebrate Zoology at UC Berkeley. Cox and his team found that the specific toxin they were measuring was hundreds, even thousands, of times more concentrated in the skins of the bats than in the seeds. These concentrations suggested that had an individual eaten one of these bats, he or she would have consumed the equivalent of more than one thousand kilograms of processed cycad flour![8] Even though an accompanying editorial remarked that such high levels of a neurotoxin in these skins might have compromised the animals' health, ensuring their fate as museum specimens, the pieces of the puzzle seemed to be coming together.

Cox and his colleagues added one final link by showing that cyanobacteria living symbiotically in the roots of the cycad trees produced large amounts of the toxin that seemed to be the culprit. So there was a complex but continuous sequence of bacteria, toxin, seeds, bats and finally humans. Fascinating stuff: a neurodegeneration caused by an unlikely, even bizarre, combination of factors. But as veterans of the aluminum hypothesis would attest, just when you think you have the case in hand, it can start to fall apart. And that's what happened with the bat idea.

First, a set of interviews with patients and controls about harvesting and eating cycad flour and/or flying foxes threw a wrench in the works by finding that only picking cycad seeds and eating the flour were risk factors, especially for men.[9] Consumption of bats appeared not to raise the risk of ALS or Parkinson's at all. Stories of young men fleeing into the woods to hide from Japanese soldiers during the war and having to eat flour made from inadequately washed cycad seeds when they were in hiding fit with the finding that men of a certain age were at greater risk. But the bat connection seemed to have been lost.

On closer examination, even the connection to picking and consuming cycad flour seemed a bit weird: risks were mixed for children, high for youths and nonexistent for adults, an inconsistent pattern that might have arisen because researchers had to rely on the answers to questionnaires given to older, sometimes demented, adults or third-party caregivers.

In the last few years, the cycad hypothesis—admittedly no longer with the help of flying foxes—has crept along, sustaining much criticism while still adding positive evidence here and there. Showing that cycad-related chemicals can induce neurodegeneration in lab animals has proven difficult; it's still not clear that whatever the Chamorros were eating could have contained concentrations of toxins high enough to cause damage (a situation not clarified by widely varying estimates of the best-candidate toxin, BMAA, beta-methylamino-L-alanine, in patients' brains). Nor are the dwindling numbers of affected people making the ongoing study any easier.

Having said that, the speculation has, if anything, grown more robust. The same Paul Cox who teamed up originally with Oliver Sacks has set out to see whether cyanobacterial blooms might be associated with neurodegenerative diseases like ALS in other parts of the world. In other words, he's asking the profound question of whether bacterial toxins might underlie neurodegenerative diseases in other parts of the world, not just in Guam. Obviously, cycads wouldn't be involved, but there are undoubtedly other ways of biomagnifying the most suspicious toxin, BMAA. In fact, in late 2013 researchers including Cox reported an unusual cluster of cases of ALS near the Thau lagoon on the Mediterranean coast, one of the hotspots for mussel and oyster consumption in France. The researchers admit they can't draw a straight line between the cases of illness found here and concentrations of toxin in these bivalves, but they wanted to put the finding on the table nonetheless.[10]

And that's where the story stands at the moment. It's reminiscent of the aluminum hypothesis I described in the last chapter.

Both are full of intriguing yet indeterminate results, both are controversial and, sadly, neither wraps up nicely with a firm conclusion. That's the way it often is.

Even though this pursuit of the curious situation in Guam might never actually result in better treatments for dementias and neurodegenerative diseases, it isn't a waste of time. The triggers for many of these conditions are unknown. (The fact that exposure to DDT has been linked to the eventual development of Alzheimer's is but one possible example.[11]) With all the (appropriate) concentration on the *genetics* of Alzheimer's disease, it's useful and important to keep open minds about other factors. You never know where knowledge will come from—sometimes it appears in the strangest places.

CHAPTER EIGHTEEN

Where You Live, What You Eat

Is it possible to reduce your chances of getting Alzheimer's? Yes, by a few per cent. And I've already mentioned some of the ways of doing this. In no particular order, you can shave off some risk points through education, a mentally challenging occupation, conscientiousness, low body weight and adequate exercise (the factor that the Ontario Brain Institute has identified as the single most effective strategy).[1] An active social life and adequate sleep are others. But it also helps to have the right genes. Of the ones you can act on, some are interrelated: education makes it easier to land a mentally stimulating job, and lower body weight is more easily maintained with a combination of exercise and healthier food. But one of the most intriguing potential links is between diet and geography. And here, the most popular example is the relationship between the Indian spice turmeric and rates of Alzheimer's in India.

The website of the integrative medicine proponent Dr. Andrew Weil features an article titled "3 Reasons to Eat Turmeric." The first, based on the research of ethnobotanist Dr. James Duke, is to alleviate Alzheimer's disease. Dr. Duke apparently found fifty studies addressing the link between turmeric (or its active chemical, curcumin) and Alzheimer's, and he connects lowered rates of Alzheimer's in India directly with turmeric's popularity there.[2]

There is a story here, but it's not as straightforward as Dr. Duke seems to suggest. Instead, it illustrates the difficulties of unambiguously identifying factors that might help people defend themselves against Alzheimer's. Starting at the chemistry end, some solid experimental data do indeed suggest that curcumin might be effective in treating the disease.[3] In the lab, curcumin both discourages amyloid beta from aggregating and begins the process of disassembling already-formed amyloid beta fibrils. Stepping up to the more complex arena of cell culture, curcumin performs well and is versatile, reversing or slowing several different chemical processes that contribute to Alzheimer's. On to whole animals: mice engineered to be susceptible to Alzheimer's were put on a relatively low curcumin diet and after six months had fewer plaques and smaller loads of plaque-building material. However, in the first of a series of puzzles, higher doses of curcumin were not protective at all.

To take into account the well-known fact that lowering plaque levels doesn't always ensure reduction of dementia, researchers tackled the cognition issue by showing that the rats which had been injected with amyloid beta but were then fed curcumin chow instead of regular lab food performed better on the Morris Water Maze test. This is a classic test of rodent cognition, one that seems a little diabolical to us humans. The setting is a two-metre-wide tank with a submerged platform in the middle. A rat is given thirty seconds to look around while on the platform and then is removed and hooded and put back in the water. The trick for the animal is to remember the way the scene looked from the perspective of the platform during training and swim back there for safety. (Neither rats, nor the more commonly used mice, are allowed to drown. If they can't find the platform, they are removed and dried.) In this case, rats suffered spatial memory impairment as a result of the injected amyloid, but the curcumin feed suppressed that memory loss.[4]

As exciting as these findings are, as studies are scaled up from cell cultures to actual human beings, definitive answers become more and more elusive. So, for instance, *Ayu*, a journal of Ayurvedic medicine (Hindu traditional medicine), published a study showing that turmeric could relieve the symptoms of Alzheimer's. The entire study was made up of *three* patients.[5] It's pretty obvious that a study of three patients doesn't really get you anywhere. You need numbers to be able to draw conclusions. Also, because the patients were all still alive, Alzheimer's would have to have been a probable, not definitive, diagnosis. Would a counterstudy of three patients who experienced no benefit allow you to conclude that turmeric was useless? No, you need numbers.

A six-month trial of thirty-four patients in Hong Kong produced some useful data about the combination of absence of side effects, levels of amyloid beta, and levels of curcumin in the blood (important because very little of it taken orally actually gets to the brain). However, because the placebo group in this experiment experienced no discernible cognitive decline over the relatively short period of the study, it was impossible to conclude anything about the efficacy of curcumin.[6]

A similar study in the U.S. had thirty-six participants: even with doses of two or four grams a day, the researchers were unable to measure significant levels of curcumin in the blood, and the patients' cognitive scores actually worsened slightly over the duration of the study. However, even these mega-doses were generally tolerated well, so it might make sense to extend similar studies to durations of years rather than months.[7]

And finally, a Singapore study with a substantial number of participants (more than one thousand) concluded that those who consumed curry with turmeric generally scored better on tests of cognitive ability than did those who rarely or never ate curry.[8] The age range was huge—sixty to ninety-three—and the group was ethnically diverse (Chinese, Indian and Malay) and chosen at a moment in time. So while the scientists tried to control for a variety of factors other than turmeric that might have influenced the results (such as cardiovascular issues, diet, exercise, consumption of alcohol and smoking), the group might well have harboured many potential influences on the results that were not discovered during the study.

For instance, Indian curries generally contain more turmeric than the Malay or Chinese versions, and they're likely consumed more often. That might explain the greater enhancement of cognitive scores among Indians in this investigation. On the other hand, Indians have a higher incidence of cardiovascular complications, which might explain why they performed worse on cognitive tests than did the Chinese. And finally, this was not an intervention to combat Alzheimer's but simply an ongoing measure of cognitive status.

How informative is this study? At best, it is suggestive, an encouragement to undertake more, and more complete, investigations. And that brings us to the claim that the low rates of Alzheimer's disease in India can be attributed to the consumption of turmeric. First, India is anything but monolithic, so it can't be assumed that the amount of turmeric consumed will be the same everywhere. The cuisines and peoples of the subcontinent (who have different genetic backgrounds) are different wherever you go. But an Indo-U.S. study begun years ago did reveal that rates of Alzheimer's were astonishingly low in Ballabgarh, a rural area about thirty kilometres from Delhi.[9] There, the rate of Alzheimer's disease for those over sixty-five was about 1 per cent, whereas in the area chosen for comparison in North America, the Monongahela Valley in Pennsylvania, that number was at least six times higher, an almost unbelievable difference.

But the researchers involved in the Indo-U.S. study did not identify turmeric as crucial in causing that difference. Food might indeed play some role, given that most Indians eat a low-fat, vegetarian diet and obesity is rare. But several other factors are likely just as important.

Not unlike the situation in North America in the 1930s, in India, dementia/Alzheimer's is taken for granted as an inevitable concomitant of aging and is likely to remain unnoticed. This is especially so because the traditional family structure is still intact, meaning that elders can count on continuing to be integrated into the family whether they are becoming demented or not. An earlier study in the Indian state of Goa reached a similar conclusion—that physicians rarely saw people with dementia because they were not seen as requiring medical intervention. In fact, there was no word in the local language, Konkani, for "dementia." (This same study also showed that family support of elders with cognitive failings was weakening, influenced more by the hope of inheritance than family fidelity.)[10]

The authors of one of the first of many reports from the Indo-U.S. study also argued that reduced life expectancy and shorter survival once dementia had set in (with little effort to ameliorate the condition medically) might have contributed to the extremely low numbers of people with dementia in the Ballabgarh study group—in effect, distinguishing it from the Pennsylvania cohort.

The cognitive testing and diagnosis carried out in India might also have overlooked many of the moderate to mild cases, especially given that existing tests had to be modified to take into account the fact that 75 per cent of the participants were illiterate. Genetics played a role too because the Ballabgarh group had a much lower incidence of the risk-inducing APOE4 gene. But even that fact couldn't even be taken at face value because, as the researchers pointed out, APOE4 contributes to heart disease too, so those carrying it might have died from *that*, never getting as far as Alzheimer's.

But Alzheimer's will inevitably be a greater problem in India in the future. There's growing evidence that low hemoglobin levels are a risk for dementia, and the Indian population tends to low hemoglobin partly because of vegetarian diets and partly because of parasites. Dr. Mary Ganguli of the University of Pittsburgh, one of the researchers most involved in the Indo-U.S. study, wrote this in an email regarding the low rates in Ballabgarh: "The long-term implication, though, is that as life expectancy improves in India and other emerging economies—the 'global aging' phenomenon—we should expect a rapid rise in the number of people with chronic disease, including dementia, in those countries. The % may still be lower than in the West, but it's a % of a much bigger denominator, so the actual numbers are also way bigger."[11]

So in reality, we should be more concerned about the overall picture in India and all developing countries, not so much whether this or that dietary component might save us all. Obviously, when it comes to curcumin, the evidence that it actually plays a significant role is pretty scant. Add to that evidence the fact that dementia rates are much higher in other parts of India than in Ballabgarh, yet those people too are consuming turmeric.

Would I nonetheless bump up my consumption of turmeric?

Why not? It appears to have few, if any, side effects, and I happen to like the cuisine that features that spice. But I wouldn't get my hopes up about its anti-Alzheimer's properties, especially because there are other, more convincing dietary steps one could take. All in all, though, great care must be taken before we conclude that diet is the primary factor in any situation.

The 2001 study comparing the Yoruba of Ibadan, Nigeria, and African-Americans in Indianapolis is a nice example.[12] The incidence (that is, the rate of appearance) of both dementia generally and Alzheimer's specifically was significantly lower in the African population: less than half the risk for both, as compared to the risk among the cohort in Indianapolis. These two populations were not too genetically dissimilar, and conditions known to predispose to Alzheimer's (diabetes, high body mass index and high blood pressure) were lower among the Africans. So this study might suggest that the westernized diet of the Americans was responsible for their greater risk. However, the African population, while genetically similar, was not identical to their American counterparts (oddly, the African population exhibited no association between the APOE4 gene and a heightened risk of Alzheimer's), so while diet was likely a factor, its importance has not been nailed down.

A similar study showing that the risk of Alzheimer's rose for Japanese people who had emigrated to Hawaii again hinted that change of diet might have played a role (more fat, less fish). However, no direct dietary data were collected on the individuals as part of that research.[13] Given that vascular issues are exacerbated by fat and that they contribute to dementia, this is a reasonable supposition, but again, we're left with inconclusiveness, if not doubts.

But thankfully, there are some definitive data about what to eat—or not—if you want to reduce your risk of dementia. One of the most persuasive studies examined the rate of cognitive decline over a six-year period in conjunction with a detailed inventory of diet and found that consumption of vegetables—*but not fruit*— reduced the risk of dementia.[14] The least cognitive decline measured over a six-year period was found in those who ate more than two vegetable portions a day on average. It was a substantial protection too, equivalent to reducing age by five years (remembering that age is the biggest risk, bar none, for Alzheimer's). Green leafy vegetables were the most effective, but the big puzzle was the lack of benefit associated with eating fruit. The authors were clearly perplexed by this finding (although it replicated a previous study, which identified green leafy and cruciferous vegetables as the most influential). They wondered whether vegetables were beneficial because they contain the most vitamin E, a powerful antioxidant. But identification of key chemical ingredients aside, this study at least identified the important food groups for reducing the risk of getting dementia.

In the midst of all this uncertainty, however, one food-related substance is unambiguously associated with Alzheimer's and dementia, and that is sugar. The connections couldn't be more pointedly described than in this comment from a 2013 research paper: "*Any* incremental increase [italics mine] in glucose levels is associated with an increased risk of dementia."[15]

Evidence has been accumulating for years that something goes wrong with the metabolism of sugar when Alzheimer's is present. One of the first measurable signals is that glucose metabolism in the brain shows signs of disruption long before any cognitive impairment can be detected. And the connection between sugar and Alzheimer's rests on insulin.

You might know the outlines of the standard story of insulin: made in the pancreas, it facilitates the uptake of sugar from the blood into recipient cells, where it's either broken down right

away and used or stored as glycogen. Disruption of the insulin/
sugar mechanism causes diabetes, of which there are two kinds,
type 1 and type 2.

Type 1 diabetes is the result of the body's immune system mistakenly attacking and destroying insulin-producing beta cells in
the pancreas, thus shutting down the production of insulin. To
help remedy this problem, insulin is injected when needed, guided
by measurements of blood sugar levels. Canadian Frederick
Banting shared the Nobel Prize for isolating and identifying insulin in the early 1920s, but much medical research has taken place
since, developing systems that deliver insulin to patients in a way
that best mimics the natural ebb and flow of the hormone.

Type 2 diabetes is a different form and accounts for more
than 90 per cent of all diabetes. It is also rapidly becoming more
common, partly, if not largely, because of high-fat and high-sugar
"Western" diets. With type 2, there's not just a shortage of insulin—although that happens. The insulin that's present in the blood
is unable to gain access to the cells that need it. This is called "insulin resistance," and the inability of tissues to take in glucose causes
a buildup of sugar in the blood and starves the tissues of energy at
the same time.

People who have type 2 diabetes have double the risk of getting Alzheimer's. In addition, glucose metabolism goes awry early
on in the disease *and* insulin receptors are most abundant in areas
of the brain like the entorhinal cortex and the hippocampus that
first fall prey to Alzheimer's. These two factors have focused the
attention of Alzheimer's researchers on the sugar-insulin relationship in the brain. Some scientists, especially Suzanne de la Monte
of Brown University, have made a powerful case—a revolutionary
case—for thinking of Alzheimer's disease as *type 3 diabetes*.[16] She
argues (as do others) that while type 2 diabetes might be a co-factor

in the development of Alzheimer's—hence the heightened risk—it doesn't cause the disease. Young mice fed an extremely high-fat diet for four months did develop type 2 diabetes and exhibited all kinds of disruptions in the brain's handling of glucose, but they showed none of the features of Alzheimer's, such as plaques and/or tangles. Conclusion: type 2 diabetes isn't sufficient to cause Alzheimer's, but there's enough overlap to suggest that they're related. (Indeed, recent experiments with mice engineered to age rapidly went one step further by finding that induced type 2 diabetes did provoke increasing amounts of amyloid and tau.)[17]

Here are some of the data which suggest that Alzheimer's is clearly a disease involving failures of the insulin system in the brain (even if the jury is still out on whether it should be called a new kind of diabetes): post-mortem Alzheimer's brains show widespread disruption of insulin production and of insulin uptake by neurons and insulin-dependent signalling pathways; animals with type 2 diabetes exhibit cognitive deficits; if normal animals are treated with a chemical that induces insulin resistance, their cognitive abilities decline, and such cognitive shortcomings can be alleviated at least partly by stimulating the brain's insulin machinery.[18]

Treating cognitive decline with insulin is not limited to lab animals. Trials involving people with mild to moderate Alzheimer's have shown that inhaling insulin can improve memory, at least in the short-term, although curiously, only for those who do not have the Alzheimer's risk gene APOE4. Those with that gene actually did worse on memory tests, suggesting once more (if we actually needed the reminder) that the situation is pretty complicated.[19] But the benefits, for some, of inhaling insulin provide more evidence of a relationship between insulin and the better-known pieces of the Alzheimer's puzzle. Another is the intimate—but destructive—relationship between insulin and amyloid beta. Any brain imbalance in insulin contributes indirectly to larger amounts of amyloid beta.

The higher-level view is simply that type 2 diabetes and its partner, obesity, are risk factors for Alzheimer's and that disturbed insulin function in the brain parallels the course of the disease, appearing early and worsening as time goes on. There is every reason to expect that if diets laden with sugar and fat are significant risk factors for type 2 diabetes, they would also be risk factors for type 3.

There might be implications for diet other than the straightforward sugar/fat connection too. Dr. de la Monte points out that a single injection of a chemical called "streptozotocin" into rats' brains causes conditions reminiscent of both Alzheimer's disease and diabetes.[20] Weeks after the injection, the rats' pancreases and blood levels of insulin were normal, but their brains were wrecked: a long list of individual pathways and molecules had been disrupted, many of these symptoms common to both diabetes and Alzheimer's, yet the damage was limited to the brain, reinforcing the idea that one kind of diabetes—this diabetes—could be brain-based.

The dietary connection here is this: the drug that accomplished all this, streptozotocin, is closely related to the nitrosamines. Nitrosamines are abundant in the foods many of us eat, including cheese, hot dogs and smoked turkey among many others. Those of us living in the West who have a thing for foods like these have been exposed to low levels of nitrosamines for many years now. To be cautious, there's no direct evidence of any contribution to the wave of Alzheimer's we're currently experiencing, but lab animals fed a diet of low levels of nitrosamines did develop both diabetes and signs of dementia.

I have not embarked on presenting the very long list of foods that are thought to offer, for one reason or another, protection against Alzheimer's. As we saw with turmeric, much of the data for such foods is fragmentary and therefore unreliable. The only advice I can give is to gather as much information as you can while evaluating each source carefully before deciding to commit to a diet involving any of these foods.

But with glucose, it's a different story. The notion that Alzheimer's might be diabetes type 3 represents a large step toward understanding the role that glucose might play in the disease. I've introduced the idea almost as an afterthought, but who knows? In the future, the diabetes connection might supersede many of the old ways of thinking about Alzheimer's. That's where we stand at the moment. So much intellectual power and money are being directed at the disease. False steps are sure to be taken; intriguing ideas are certain to be jettisoned when it becomes clear that they're impractical. But some day enough pieces of information will be gathered to make inroads into the disease.

CHAPTER NINETEEN

What's Next?

The science of Alzheimer's is sometimes bewilderingly complex. But given that its realm is the human brain and that the
disease takes decades to manifest itself, this shouldn't really
come as a surprise. For instance, we have nothing like a complete
picture of how memories are formed and retained in the brain, so
the attempt to understand how they fail in Alzheimer's is already
handicapped. And even though Alzheimer himself lit the fuse
on this research more than a hundred years ago, it's true that the
intensive search for a treatment and cure (but first for understanding!) has been flourishing only since the mid-1970s. For an intricate piece of science, that is not a long time.

There is encouraging news, however: we're much further ahead
than we were in the 1970s. The roles of amyloid beta plaques and
tau-based tangles are much clearer (if still somewhat uncertain); the
categorization of the various dementias, including Alzheimer's, is
much more advanced; and the ideas for treatments are based on a
much broader platform of science than ever before.

But even these examples of progress highlight the frustrations.

Yes, plaques and tangles have dominated the discussion since Alois Alzheimer took notice of them. But are they no. 1 and no. 2 or no. 1 and no. 1 (a) or . . . ? Most researchers would tilt toward plaques, or at least the amyloid beta that is their main ingredient, as the essential cause, but Alzheimer's scientists are by no means unanimous in their support of this hypothesis. To me, there are other fascinating possibilities. For one thing, the accumulation of tangles parallels the spread of the disease in the brain, whereas plaques first appear in the default network (in neurons an appreciable distance from the hippocampus and entorhinal cortex, where the disease really takes shape). And what about the fact that those entorhinal and hippocampal neurons, which are the last to be wrapped with myelin as the brain develops, are also the first to crash?

Understanding the connection between plaques, tangles and the predecessors of both is one of the most important goals of Alzheimer's research today, although in the end that understanding might need to be based on other approaches, such as seeing Alzheimer's as type 3 diabetes. Such an understanding would be of great satisfaction to researchers, not just because it would untie some theoretical knots, but also because it would make the path to new treatments much clearer.

Some think the obsession with plaque-reducing or plaque-preventing drugs is a mistake, notwithstanding the fact that drug companies have steamed ahead, pledging hundreds of millions of dollars of their own (and shareholders') money to that very end. The prize—the effective drug—would obviously be of immeasurable value if the patient numbers continue to grow as they are at present. Like so many others, I have a personal interest in this quest: I'd rather have an effective drug sooner than later, given that I've entered the susceptible years.

On good days, I add up the things that I think should help protect me from Alzheimer's: good education (thanks, Winnipeg!), challenging jobs, not too overweight, in reasonable and relatively stable physical condition, conscientious, and maybe most important of all, a big head. (As opposed to a swelled head, which is no help at all!)

But I don't know the flip side of the coin. Do I have one or even two APOE4 genes? After all, my mother had dementia when she died and could have had one or two of those genes (or none, given that we never knew whether she had Alzheimer's or a different kind of dementia). My dad was cognitively intact until the end, so I'd guess (and it *is* a guess) that he had had one APOE4 gene at most. But even if I had one, knowing that the increase in risk for men with the gene is much less than for women, I would revert to thinking that the protective factors I cling to will see me through.

But isn't that one of the most arresting things about this disease? That behaviours, like sticking with school, eating well, playing ball or music, must have some sort of concrete effect on the brain: some new dendritic sprouts, a boost in a set of neurotransmitters or ramped-up insulin activity and glucose processing. The immaterial, like words and thoughts, can reach across the gap and make material change in the brain, much of which is protective. We can explain this phenomenon by arguing that the more the brain is used, the more likely it will generate new synapses or even new neurons, but that bland explanation doesn't get at the fine, sure-to-be-fascinating details.

And the history of Alzheimer's is so rich. Even though the pace of Alzheimer's research has intensified dramatically over the last forty years, its beginnings were dynamic. At the turn of the last century, a uniquely creative group of German and Czech neurologists, pathologists and microscopists first identified the signs of a variety of neuropathologies that still plague us today:

Alois Alzheimer (whose plaques and tangles were made visible by Bielschowsky's and Nissl's stains), Frederic Lewy (who discovered the Lewy bodies associated with Parkinson's and dementia), Arnold Pick (the head of the Prague neuropathological school, who found the abnormal amounts of tau that came to be known as "Pick bodies") and both Hans Gerhard Creutzfeldt and Alfons Maria Jakob (who identified Creutzfeldt-Jakob disease). It was a time of fantastic creativity and tough competition, and we are still benefiting from their pioneering work—not the least of which was the mid-1990s conclusion, based on Dr. Alzheimer's rediscovered slides, that Auguste Deter had early-onset Alzheimer's disease triggered by a mutation in the gene presenilin 1.[*]

As is the case with any science, a number of personalities stand out over the last century, for varied reasons: Auguste Deter herself and her anguish revealed in the interviews with Alois Alzheimer; William Osler and his off-the-cuff joke about getting rid of people over sixty-five; Donald Crapper McLachlan and his pursuit of the effects of aluminum; Sister Mary and her amazing, plaque-ridden but totally functional brain; Jonathan Swift; Rita Hayworth. They've come from all walks of life.

Since age is the greatest risk factor for Alzheimer's, scientists looking for remedies and a cure have the great advantage of widening their research to include studying the secrets of healthy aging. And at least in the most economically advantaged countries, people are living longer lives, with a startling and steady increase in life expectancy since the mid-eighteen hundreds: a month increase for every four that pass—that is the definition of consistency.[1] Those statistics have provoked the claim that we're no longer bound by the venerable idea that the human life span is

---

[*] Now somewhat tentative.

fixed but life expectancy varies. Is it really possible that the human life span, the maximum age to which we *can* live, is already being extended? And what if greater understanding of the factors that cause us to age (like shortening telomeres) will permit us to tinker and extend our life span even further? Of course, that thinking then rebounds quickly to wondering whether an extended life is desirable unless dementia is conquered.

One of the key issues in the search for a treatment or cure for Alzheimer's is "brain reserve" (those protective factors that I mentioned earlier in this chapter) and general health improvements that might obliquely reduce the risk of dementia. The early results of studies in Europe showing that the rate of dementia is falling must be replicated over and over again to ensure that they are (1) real and (2) robust in the long term.[2] The authors of the U.K. study showing a dramatic decline in the prevalence of dementia suggested that education and better cardiovascular health are at the root of it, and that possibility is encouraging. But it's by no means certain. The rising tide of obesity and diabetes threatens to undermine such advances.

What can a person do to ensure a long, healthy and cognitively intact life? Of course, you can do nothing about the genes you've been dealt, but when it comes to late-onset Alzheimer's, those genes do not guarantee inevitability; they increase risk, and risk can be modified. If you're starting to think about this in your thirties or forties, some things, like childhood nutrition and education, are also beyond your reach. However, as the Ontario Brain Institute has declared, daily exercise is the first and more important thing on the list, and that can be exercise that is threaded through your everyday life, like walking. Ensuring you are not overweight is another. The weight issue can be addressed partly by choosing the right low-fat, low-sugar, high-anti-oxidant foods.

While you might already be doing what you can to stave off dementia, you can be sure that scientists will be working flat out, with the goal of understanding the disease well enough to develop effective treatments. Some themes will surely dominate. One is the idea that Alzheimer's begins years, even decades, before symptoms appear. That principle will likely guide most of the upcoming clinical trials. From which direction might those treatments come? We will know in a few years whether the amyloid-reducing approach being applied to the at-risk extended family in Colombia actually works—for them. The next step would then be to see whether this approach applies more generally. In addition, several companies are working on the development of new antibodies designed to act early in the disease. There are also plans to target tangles rather than plaques, and the outcome of such efforts might settle once and for all which of the two is the instigator.

Efforts to reduce the disarray of insulin production and signalling in the brain will likely be a third direction, and there will inevitably be others that, at this point, are too difficult to predict. One thing is sadly sure, though: the timeline for the development and approval of new drugs is substantial, years and years. That reality says to me that taking steps to limit the risk factors that are already well known (see above) is not just prudent; it's really the only thing you can do.

That's a little bit bleak, but after all, we're talking about a disease of prodigious proportions, far beyond our ability to control, at least for now. On the other hand, the pace of the science today is fantastic. While I was writing this book, two new blood tests for Alzheimer's, or at least tests to foresee the development of the disease, were announced: one based on a set of ten fat molecules in the blood, the other on a set of ten proteins. In both cases, the combination of results from all ten could predict progression to

Alzheimer's over the next few years with 80 to 90 per cent accuracy. Of course, the question arises: Who would want to know and who would rather not?[3]

When it comes to treatment, there are already trials going on, like the one in Colombia, that will tell us something crucial. But one of the key issues in Alzheimer's research is this: How many disappointing results can you endure as you're waiting for the sensationally great one? And how do you choose which, out of many promising, high-quality lines of research, is the one that will take you furthest down the road to a treatment for the disease? Once you make that choice, you have to fund it somehow. That's what's happening right now, and it's powering a lot of science. Some of it will pay off.

I'd be excited to live long enough to see the whole thing resolved. It will be one day, but that day is not exactly on the horizon.

Acknowledgements

Writing can look like—and certainly feel like—a lonely business, but for this book, there are many people who have helped guide me.

First, my interest in Alzheimer's was originally sparked by cases of the disease in my family. You cannot go face to face with the disease without wondering what's going on. It's that difficult side of medicine: we learn from those afflicted.

I am fortunate to know several prion researchers whose work, if not directly aimed at Alzheimer's, has shed valuable light on the disease. Neil Cashman, Valerie Sim, Stefanie Szub and Lary Walker are those I know best. In addition, several researchers who address Alzheimer's directly responded quickly and graciously to questions I forwarded to them: Ellen Bialystok, Mary Ganguli, Cheryl Grady, Manuel Graeber and Derek Lowe are notable examples.

There are also people with whom I talk science who undoubtedly influence me even more than I realize—people like Trevor Day, Judy Illes, Kevin Keough, Christie Nicholson, Penny Park, John Rennie and Sam Weiss. I enjoyed watching Jeff Joseph dissect a couple of human brains at Foothills Hospital in Calgary (one shrunken by Alzheimer's), and Rosie Redfield's blog alerted me to the Lejeune affair. Finally, the 140 or so alumni of the Banff Science Communications Program make those two weeks my favourite part of the summer. Half a month talking about science and communication and eating bacon—what could be better? (Although bacon and the risk of Alzheimer's? Hmmmm.)

Having access to the University of Toronto library online makes thorough research dramatically easier than it used to be. It still amazes me that I can download scientific accounts from a hundred or even two hundred years ago—in seconds.

As I say, authors never work alone. In my case, I have stalwart help. Jim Gifford has shepherded this book through from concept to final product, and my agent, Jackie Kaiser, is hugely supportive. I realize that in a previous set of acknowledgements I said that what made her happiest was that I had finished the book. I suspect that's true this time around as well.

I acknowledge the patience of many friends, who had to sit listening as I rambled on about the latest most intriguing thing I'd just discovered about Alzheimer's disease. Unfortunately, that's not likely to stop soon. I have spared Rachel, Amelia and Max from much of that, but I will likely demand that they read this book. Mary Anne Moser, my partner in things, was also consistent in her desire to read the manuscript despite my bending her ear all the way along. Now that's dedication.

Notes

Introduction

1 http://report.nih.gov/categorical_spending.aspx.
2 R.B. Lipton et al., "Exceptional Parental Longevity Associated with Lower Risk of Alzheimer's Disease and Memory Decline," *Journal of the American Geriatrics Society* 58, no. 6 (June 2010): 1043–49.

Chapter 1: Facing, or Fearing, Aging

1 http://penelope.uchicago.edu/Thayer/E/Roman/Texts/Ptolemy/Tetrabiblos/4C* .html#9.
2 Ibid., 207.
3 www.explorethomascole.org/tour/items/75/series.
4 Nathanael Emmons, "Shortening of Life" (Sermon VII) in *The Works of Nathanael Emmons; with a Memoir of His Life,* ed. Jacob Ide, vol. 3 (Boston: Crocker & Brewster, 1842), 82.
5 William A. Alcott, *Laws of Health* (Boston: John P. Jewett, 1857), 9, http://catalog. hathitrust.org/Record/010600732.
6 George Miller Beard, *Legal Responsibility in Old Age: Based on Researches into the Relation of Age to Work: Read before the Medico-legal Society of the City of New York at the Regular Meeting of the Society, March, 1873* . . . (New York: Russells' American Steam Printing House, 1874).
7 Ibid., 5.
8 W. Osler, "Valedictory Address at Johns Hopkins University," *Journal of the American Medical Association* 44 (1905): 705–10.

9 Ibid., 707.

10 Ibid., 708.

11 Ernest M. Gruenberg, "The Failures of Success," *Milbank Quarterly* 55, no. 1 (March 1977): 3–24.

12 http://www.alz.org/downloads/facts_figures_2012.pdf.

Chapter 2: "I have, so to say, lost myself"

1 Konrad Maurer, Stephan Volk, and Hector Gerbaldo, "Auguste D. and Alzheimer's Disease," *Lancet* 349 (1997): 1546–49.

2 G. Cipriani et al., "Alzheimer and His Disease: A Brief History," *Neurological Science* 32 (2011): 275–79.

3 R. Dahm, "Alzheimer's Discovery," *Current Biology* 16, no. 21 (2006): R906–R910.

4 Manuel Graeber, "No Man Alone: The Rediscovery of Alois Alzheimer's Original Cases," *Brain Pathology* 9 (1999): 237–40.

5 Peter J. Whitehouse, *The Myth of Alzheimer's* (New York: St. Martin's Griffin, 2008), 78.

6 Manuel Graeber, "Reanalysis of the First Case of Alzheimer's Disease," *European Archives of Psychiatry and Clinical Neurosciences* 249 (1999), supplement 3, III/10–III/13.

7 A. Alzheimer, "Über eigenartige Krankheitsfälle des späteren," *Alters. Zbl Ges Neurol Psych* 4 (1911): 356–85 (translated by Taggart Wilson, 2014).

Chapter 3: Has Alzheimer's Always Been with Us?

1 G.E. Berrios, "Alzheimer's Disease: A Conceptual History," *International Journal of Geriatric Psychiatry* 5 (1990): 355–65.

2 "On Old Age," section 7, www.gutenburg.org/files/2808/2808=h/2808=h.htm.

3 Jean-Étienne Dominique Esquirol, "Idiocy," in *Mental Maladies: A Treatise on Insanity* (Philadelphia: Lee and Blanchard, 1845), 447.

4 Ibid., 418.

5 Ibid.

6 Thomas Jameson, *Essays on the Changes of the Human Body at Its Different Ages* (London: Longman, Hurst, Rees, Orme, and Brown, 1811), 138–39.

7 Robert Katzman and Katherine Bick, eds., *Alzheimer Disease: The Changing View* (San Diego: Academic Press, 2000).

8 J.M.M. Lage, "100 Years of Alzheimer's Disease (1906–2006)," *Journal of Alzheimer's Disease* 9 (2006): 15–26.

9 D. Rothschild and M.A. Trainor, "Pathologic Changes in Senile Psychoses and Their Psychobiologic Significance," *American Journal of Psychiatry* 93 (1937): 757–88.

10 David C. Wilson, "The Pathology of Senility," *American Journal of Psychiatry* 111 (1955): 902–6.

11 David Stonecypher, "Old Age Need Not Be Old," *New York Times Magazine,* August 18, 1957, 27ff.

12 Walter Alvarez, "Cerebral Arteriosclerosis with Small, Commonly Unrecognized Apoplexies," *Geriatrics* 1 (1946): 189–216.

13 Robert Katzman, "The Prevalence and Malignancy of Alzheimer Disease," *Archives of Neurology* 33 (April 1976): 217–18.

14 Katzman and Bick, *Alzheimer Disease,* 266.

15 http://blog.alz.org/dear-abby-voice-for-alzheimers/.

Chapter 4: The Case of Jonathan Swift

1 Jonathan Swift, *Gulliver's Travels,* ed. Colin McKelvie. (Belfast: Appletree Press, 1976), 182–83.

2 Ibid., 185.

3 Ibid.

4 Ibid.

5 James C. Harris, "Gulliver's Travels: The Struldbruggs," *Archives of General Psychiatry* 62 (March 2005): 243–44.

6 J. Banks, "The Writ 'de Lunatico Inquirendo' (Swift)," *Dublin Quarterly Journal of Medical Science* 31 (1861).

7 P. Crichton, "Jonathan Swift and Alzheimer's Disease," *Lancet* 342 (1993): 874.

8 W.R. Brain, "The Illness of Dean Swift," *Irish Journal of Medical Science* (1952): 320–21, 337–46.

9 J. Houston, "Phrenological Report on Jonathan Swift's Skull," *Phrenological Journal* 9 (1834/6): 558–60.

Chapter 5: The Biology of Aging

1 S.G. Lenhoff and H.M. Lenhoff, *Hydra and the Birth of Experimental Biology, 1744* (Pacific Grove, CA: Boxwood Press, 1986).

2 Lenhoff and Lenhoff, "First Memoir," in *Hydra,* 8.

3 Anna-Marie Boehm et al., "FoxO Is a Critical Regulator of Stem Cell Maintenance in Immortal Hydra," *Proceedings of the National Academy of Sciences* 109, no. 48 (November 2012): 19697–702.

4 D. Chen et al., "Germline Signaling Mediates the Synergistically Prolonged Longevity Produced by Double Mutations in daf-2 and rsks-1 in *C. elegans,*" *Cell Reports* 5 (December 26, 2013): 1600–1610.

5 D. Dubai et al., "Life Extension Factor Klotho Enhances Cognition," *Cell Reports* 7, no. 4 (May 2014): 1065–76.

6 C.A. Stephens, *Salvation by Science (Natural Salvation)* (Norway Lake, ME: The Laboratory, 1913), 14.

7 Ibid., 29.

8 A.H. Ebeling, "Dr. Carrel's Immortal Chicken Heart," *Scientific American* 166 (1942): 22–24.

9 http://en.wikipedia.org/wiki/List_of_living_supercentenarians.

10 Aubrey D.N.J. de Grey, "Biogerontologists' Duty to Discuss Timescales Publicly," *Annals of the New York Academy of Sciences* 1019 (2004): 542–45.

11 A. Olovnikov, "Telomeres, Telomerase and Aging: Origin of the Theory," *Experimental Gerontology* 31, no. 4 (1996): 443–48.

12 J.A. Mattison et al., "Impact of Caloric Restriction on Health and Survival in Rhesus Monkeys from the NIA Study," *Nature* 489 (2012): 318–21.

13 Ricki J. Colman et al., "Caloric Restriction Reduces Age-Related and All-Cause Mortality in Rhesus Monkeys," *Nature Communications* 5, article 3557 (April 1, 2014).

14 Leonard Hayflick, "How and Why We Age," *Experimental Gerontology* 33, no. 7/8 (1998): 639–53.

CHAPTER 6: A NATURAL LIFE

1 Edmond Halley, *An Estimate of the Degrees of the Mortality of Mankind,* http://www.pierre-marteau.com/editions/1693-mortality.html.

2 Postscript, ibid., 654.

3 Jake Edmiston, "Dead Men Walking: Under 19th-Century Conditions, Millions of Canadians Would Already Be Dead," *National Post*, October 26, 2013, with statistics from Statistics Canada.

4 O. Burger, A. Baudisch, and J.W. Vaupel, "Human Mortality Improvement in Evolutionary Context," *Proceedings of the National Academy of Sciences of the United States of America* 109, no. 44 (2012): 18210–214.

5 Ibid., 18211.

6 G.C. Myers and K.G. Manton, "Compression of Mortality: Myth or Reality?" *Gerontologist* 24, no. 4 (1984): 346–53.

7 L.I. Dublin, *Health and Wealth* (New York: Harper, 1928).

8 J.F. Fries, "Aging, Natural Death and the Compression of Morbidity," *New England Journal of Medicine* 303, no. 3 (1980): 130–35.

9 Ibid., 130.

10 J.F. Fries, B. Bruce, and E. Chakravarty, "Compression of Morbidity 1980–2011: A Focused Review of Paradigms and Progress," *Journal of Aging Research*, Article ID 261702, 2011 (2011), http://dx.doi.org/10.4061/2011/261702.

11 J. Robine et al., "Is There a Limit to the Compression of Mortality?" presented at the Living to 100 and Beyond Symposium, Orlando, FL, January 7–9, 2008.

12 H. Strulik and S. Vollmer, "Long-Run Trends of Human Aging and Longevity," *Program on the Global Demography of Aging*, Working Paper 73 (August 2011).

13 K. Christensen et al., "Ageing Populations: The Challenges Ahead," *Lancet* 374 (2009): 1196–1208.

14 http://www.usda.gov/factbook/chapter2.pdf. See also K.M. Flegal et al., "Prevalence of Obesity and Trends in the Distribution of Body Mass Index among U.S. Adults, 1999–2010," *Journal of the American Medical Association* 307, no. 5 (2012): 491–97.

15 Flegal et al., "Prevalence of Obesity," 491–97.

16 L.K. Twells et al., "Current and Predicted Prevalence of Obesity in Canada: A Trend Analysis," *Canadian Medical Association Journal Open* 2, no. 1 (March 3, 2014), E18–E26.

17 J. Olshansky et al., "A Potential Decline in Life Expectancy in the United States in the 21st Century," *New England Journal of Medicine* 352, no. 11 (March 17, 2005): 1138–45.

18 K.M. Flegal et al., "Association of All-Cause Mortality with Overweight and Obesity Using Standard Body Mass Index Categories," *Journal of the American Medical Association* 309, no. 1 (January 2, 2013): 71–82.

CHAPTER 7: THE AGING BRAIN

1 R. Epstein, "Brutal Truths about the Aging Brain," *Discover* 33, no. 8 (2012): 48.

2 G.A. Miller, "The Magical Number Seven, Plus or Minus Two: Some Limits on Our Capacity for Processing Information," *Psychological Review* 63 (1956): 81–97.

3 E. Tulving, "Episodic Memory: From Mind to Brain," *Annual Review of Psychology* 53 (2002): 1–25.

4 S. Corkin, "Lasting Consequences of Bilateral Medial Temporal Lobectomy: Clinical Course and Experimental Findings in H.M.," *Seminars in Neurology* 4, no. 4 (1984): 249–59.

5 M. St-Laurent et al., "Influence of Aging on the Neural Correlates of Autobiographical, Episodic and Semantic Memory Retrieval," *Journal of Cognitive Neuroscience* 23, no. 12 (2011): 4150–63.

6 G. Yang et al., "Stably Maintained Dendritic Spines Are Associated with Lifelong Memories," *Nature* 462 (December 2009): 920–25.

CHAPTER 8: PLAQUES AND TANGLES

1 H.A. Archer et al., "Amyloid Load and Cerebral Atrophy in Alzheimer's Disease: An 11C-PIB Positron Emission Tomography Study," *Annals of Neurology* 60, no. 1 (July 2006): 145–47.

2 P.V. Arrigata et al., "Neurofibrillary Tangles But Not Senile Plaques Parallel Duration and Severity of Alzheimer's Disease," *Neurology* 42, no. 3 (March 1991): 631–39.

3 R.L. Buckner et al., "Molecular, Structural, and Functional Characterization of Alzheimer's Evidence for a Relationship between Default Activity, Amyloid, and Memory Disease," *Journal of Neuroscience* 25, no. 34 (2005): 7709–17.

4 L.M. Ittner and J.L. Gotz, "Amyloid-β and Tau: A Toxic *Pas de Deux* in Alzheimer's Disease," *Nature Reviews Neuroscience* 12, no. 2 (February 2011): 65–72.

5 T. Bolmont et al., "Induction of Tau Pathology by Intracerebral Infusion of Amyloid-Beta-Containing Brain Extract and by Amyloid-Beta Deposition in APP and Tau Transgenic Mice," *American Journal of Pathology* 171 (2007): 2012–20.

CHAPTER 9: "I ONLY RETIRE AT NIGHT"

1 David Snowdon, "Aging and Alzheimer's Disease: Lessons from the Nun Study," *Gerontologist* 37, no. 2 (1997): 150–56.
2 D. Snowdon et al., "Linguistic Ability in Early Life and Cognitive Function and Alzheimer's Disease in Late Life," *Journal of the American Medical Association* 275, no. 7 (February 21, 1996): 528–32.
3 S. Kemper et al., "Language Decline across the Life Span," *Psychology and Aging* 16, no. 2 (2001): 227–39.
4 D.D. Danner et al., "Positive Emotions in Early Life and Longevity: Findings from the Nun Study," *Journal of Personality and Social Psychology* 80, no. 5 (2001): 804–13.
5 R.S. Wilson et al., "Religious Orders Study: Overview and Change in Cognitive and Motor Speed," *Aging, Neuropsychology, and Cognition* 11, nos. 2 & 3 (2004): 280–303.
6 D.A. Bennett et al., "Education Modifies the Relation of AD Pathology to Level of Cognitive Function in Older Persons," *Neurology* 60 (2003): 1913.

CHAPTER 10: A DEADLY PROGRESSION

1 H. Braak and K. Del Tredici, "The Pathological Process Underlying Alzheimer's Disease in Individuals under Thirty," *Acta Neuropathologica* 121 (2011):171–81.
2 H. Braak and E. Braak, "Neuropathological Stageing of Alzheimer-Related Changes," *Acta Neuropathologica* 82 (1991): 239–59.
3 C.R. Jack et al., "Tracking Pathophysiological Processes in Alzheimer's Disease: An Updated Hypothetical Model of Dynamic Biomarkers," *Lancet Neurology* 12 (February 2013): 207–16.
4 T. Gomez-Isla et al., "Profound Loss of Layer II Entorhinal Cortex Neurons Occurs in Very Mild Alzheimer's Disease," *Journal of Neuroscience* 16, no. 14 (July 15, 1996): 4491–4500.
5 H. Braak and E. Braak, "Development of Alzheimer-Related Neurofibrillary Changes in the Neocortex Inversely Recapitulates Cortical Myelogenesis," *Acta Neuropathologica* 92 (1996): 197–201.
6 W. Huijbers et al., "Amyloid Deposition Is Linked to Aberrant Entorhinal Activity among Cognitively Normal Older Adults," *Journal of Neuroscience* 34, no. 15 (April 2014): 5200–5210.
7 M. Meyer-Luehmann et al., "Exogenous Induction of Cerebral Beta-Amyloido-Genesis Is Governed by Agent and Host," *Science* 313 (2006):1781–84.
8 R. Morales et al., "*De novo* Induction of Amyloid-B Deposition *in Vivo*," *Molecular Psychiatry* 17 (2012): 1347–53.

9 J. Stohr et al., "Purified and Synthetic Alzheimer's Amyloid Beta (Aβ) Prions," *Proceedings of the National Academy of Sciences* 109, no. 27 (July 3, 2014): 11025–30.

10 J. Hardy and T. Revesz, "The Spread of Neurodegenerative Disease," *New England Journal of Medicine* 366, no. 22 (May 31, 2012): 2126–28.

11 S. Nath et al., "Spreading of Neurodegenerative Pathology via Neuron-to-Neuron Transmission of β-Amyloid," *Journal of Neuroscience* 32, no. 26 (2013): 8767–77.

12 J.J. Stamps et al., "A Brief Olfactory Test for Alzheimer's Disease," *Journal of the Neurological Sciences* 333 (2013): 19–24.

Chapter 11: The Brain Fights Back

1 R. Katzman, "Clinical, Pathological and Neurochemical Changes in Dementia: A Subgroup with Preserved Mental Status and Numerous Neocortical Plaques," *Annals of Neurology* 23 (1988): 138–44.

2 P.W. Schofield et al., "An Association between Circumference and Alzheimer's Disease in a Population-Based Study of Aging and Dementia," *Neurology* 49 (1997): 30–37.

3 N.A. Royle et al., "Estimated Maximal and Current Brain Volume Predict Cognitive Ability in Old Age," *Neurobiology of Aging* 34 (2013): 2726–33.

4 J.B. Jorgensen et al., "The Correlation between External Cranial Volume and Brain Volume," *American Journal of Physical Anthropology* 19, no. 4 (December 1961): 317–20.

5 *Education at a Glance Highlights* (OECD, 2012).

6 M. Zhang et al., "The Prevalence of Dementia and Alzheimer's Disease in Shanghai, China: Impact of Age, Gender, and Education," *Annals of Neurology* 27 (1990): 428–37.

7 Yaakov Stern, "Cognitive Reserve in Ageing and Alzheimer's Disease," *Lancet Neurology* 11 (2012): 1006–12.

8 T.A. Schweizer et al., "Bilingualism as a Contributor to Cognitive Reserve: Evidence from Brain Atrophy in Alzheimer's Disease," *Cortex* 48 (2012): 991–96.

9 B.T. Gold et al., "Lifelong Bilingualism Maintains Neural Efficiency for Cognitive Control in Aging," *Journal of Neuroscience* 33, no. 2 (January 2013): 387–96.

10 R.S. Wilson et al., "Conscientiousness and the Incidence of Alzheimer Disease and Mild Cognitive Impairment," *Archives of General Psychiatry* 64, no. 10 (2007): 1204–12.

11 E. Neuvonen et al., "Late-Life Cynical Distrust, Risk of Incident Dementia, and Mortality in a Population-Based Cohort," *Neurology*, published online before print, May 28, 2014.

12 V.M. Moceri et al., "Using Census Data and Birth Certificates to Reconstruct the Early-Life Socioeconomic Environment and the Relation to the Development of Alzheimer's Disease," *Epidemiology* 12, no. 4 (July 2001): 383–89.

13 H.A. Lindstrom, "The Relationships between Television Viewing in Mid-life and the Development of Alzheimer's Disease in a Case-Control Study," *Brain and Cognition* 58 (2005): 157–65.

14 R.S. Wilson et al., "Participation in Cognitively Stimulating Activities and Risk of Incident Alzheimer Disease," *Journal of the American Medical Association* 287 (2002): 742–48.

15 P. Vemuri et al., "Association of Lifetime Intellectual Enrichment with Cognitive Decline in the Older Population," *JAMA Neurology* 71, (June 23, 2014); M. Wirth et al., "Gene-Environment Interactions: Lifetime Cognitive Activity, APOE Genotype, and β-Amyloid Burden," *Journal of Neuroscience* 34, no. 25 (2014).

16 G. Kempermann, "What the Bomb Said about the Brain," *Science* 340, no. 1180 (2013).

17 J. Freund et al., "Emergence of Individuality in Genetically Identical Mice," *Science* 340 (May 2013): 756–59.

18 E.A. Maguire et al., "Navigation-Related Structural Change in the Hippocampi of Taxi Drivers," *Proceedings of the National Academy of Sciences* 97, no. 8 (February 11, 2000): 4398–4403.

CHAPTER 12: IS THE EPIDEMIC SLOWING?

1 F.E. Matthews et al., "A Two-Decade Comparison of Prevalence of Dementia in Individuals Aged 65 Years and Older from Three Geographical Areas of England: Results of the Cognitive Function and Ageing Study I and II," *Lancet* 382, no. 9902 (October 26, 2013): 1405–12.

2 K.C. Manton et al., "Declining Prevalence of Dementia in the U.S. Elderly Population," *Advances in Gerontology* 16 (2005): 30–37.

3 E.M.C. Schrijvers et al., "Is Dementia Incidence Declining? Trends in Dementia Incidence since 1990 in the Rotterdam Study," *Neurology* 78 (May 2012): 1456–63.

4 James Flynn, "Beyond the Flynn Effect: Solution to All Outstanding Problems—Except Enhancing Wisdom" (lecture, University of Cambridge), http://www.psychometrics.cam.ac.uk/about-us/directory/beyond-the-flynn-effect.

5 T.M. Hughes et al., "Arterial Stiffness and β-Amyloid Progression in Non-demented Elderly Adults," *JAMA Neurology* 71 (March 31, 2014): 562–68.

6 A.J. Schneider et al., "The Neuropathology of Probable Alzheimer's Disease and Mild Cognitive Impairment," *Annals of Neurology* 66, no. 2 (August 2009): 200–208.

7 J. Joseph et al., "Copernicus Revisited: Amyloid Beta in Alzheimer's Disease," *Neurobiology of Aging* 22 (2001): 131–46.

8 R.J. Castellani, "Reexamining Alzheimer's Disease: Evidence for a Protective Role for Amyloid-β Protein Precursor and Amyloid-β," *Journal of Alzheimer's Disease* 18 (2009): 447–52.

9 Peter J. Whitehouse, *The Myth of Alzheimer's* (New York: St. Martin's Griffin, 2008), 78, 56.

10 R. Katzman, "The Prevalence and Malignancy of Alzheimer Disease," *Archives of Neurology* 33 (April 1976): 217–18.

11 Castellani, "Reexamining Alzheimer's Disease," 448.

Chapter 13: Am I going to get it? And if so, when?

1   G.A. Jervis, "Early Senile Dementia in Mongoloid Idiocy," *American Journal of Psychiatry* 105 (1948): 102–6.

2   M. Gautier and P.S. Harper, "Fiftieth Anniversary of Trisomy 21: Returning to a Discovery," *Human Genetics* 126 (2009): 317–24.

3   Barbara Casassus, "Down's Syndrome Discovery Dispute Resurfaces in France," *Nature* (February 11, 2014).

4   L. Heston, "Alzheimer's Disease, Trisomy 21 and Myeloproliferative Disorders: Associations Suggesting a Genetic Diathesis," *Science* 196 (1977): 322–23.

5   There's no space here either to expand on the intricate detail underlying the mechanics of genetics as it is understood today or to follow the twisting, turning, often frustrating scientific pursuit of those details. But geneticist Rudy Tanzi's book *Decoding Darkness*, a personal account of his involvement in the science, gives a good sense of both. See Tanzi and A.B. Parson, *Decoding Darkness* (New York: Basic Books, 2000).

6   http://www.ncbi.nlm.nih.gov/books/NBK1236/.

7   Ibid.

8   L. Wu et al, "Early-Onset Familial Alzheimer's Disease (EOFAD)," *Canadian Journal of Neurological Sciences* 39 (2012): 436–45.

9   B. De Strooper and T. Voet, "A Protective Mutation," *Nature News and Views* 488 (August 2, 2012): 38–39.

10 http://www.ncbi.nlm.nih.gov/books/NBK1236/.

11 L. Bertram and R.E. Tanzi, "The Genetics of Alzheimer's Disease," *Progress in Molecular Biology and Translational Science* 107 (2012): 79–101.

12 Margaret Lock, *The Alzheimer Conundrum* (Princeton, NJ: Princeton University Press, 2013), 144.

13 A. Altmann et al., "Sex Modifies the APOE-Related Risk of Developing Alzheimer Disease," *Annals of Neurology* 14 (April 2014) (early view).

14 D.R. Nyholt et al., "On Jim Watson's APOE Status: Genetic Information Is Hard to Hide," *European Journal of Human Genetics* 17, no. 2 (February 2009): 147–49.

15 C.E. Finch and C.B. Stanford, "Meat-Adaptive Genes and the Evolution of Slower Aging in Humans," *Quarterly Review of Biology* 79, no. 1 (March 2004): 3–50.

16 A.B. Graves et al., "The Association between Head Trauma and Alzheimer's Disease," *American Journal of Epidemiology* 131, no. 3 (1990): 491–501; A.P. Spira et al., "Self-Reported Sleep and β-Amyloid Deposition in Community-Dwelling Older Adults," *JAMA Neurology* 70 (December 1, 2013), 1537–43; F. Sztark et al., "Exposure to General Anaesthesia Could Increase the Risk of Dementia in Elderly: 18AP1-4," *European Journal of Anaesthesiology* 30 (June 2013): 245.

Chapter 14: Candidates But No Champions

1 Dale Schenk et al., "Immunization with Amyloid-β Attenuates Alzheimer-Disease-Like Pathology in the PDAPP Mouse," *Nature* 400 (July 1999): 173–77.

2 B. Vellas et al., "Long-Term Follow-Up of Patients Immunized with AN1792: Reduced Functional Decline in Antibody Responders," *Current Alzheimer's Research* 6, no. 2 (April 2009): 144–51.

3 http://pipeline.corante.com/archives/alzheimers_disease/

4 A.D. Watt et al., "Do Current Therapeutic Anti-Aβ Antibodies for Alzheimer's Disease Engage the Target?" *Acta Neuropathologica* (May 2014) (advance online publication).

5 "Colombia at the Centre of Preclinical AD Research," *The Lancet Neurology* 11, no. 7 (July 2012): 567.

6 A.S. Fleisher et al., "Florbetapir PET Analysis of Amyloid-β Deposition in the Presenilin 1 E280A Autosomal Dominant Alzheimer's Disease Kindred: A Cross-Sectional Study," *Lancet Neurology* 11 (2012): 1057–65.

7 M. Mapstone et al., "Plasma Phospholipids Identify Antecedent Memory Impairment in Older Adults," *Nature Medicine* (March 9, 2014) (advance online publication).

8 A. Lozano, "Tuning the Brain," *The Scientist* (October 28, 2013), www.the-scientist.com.

9 Y. Sheline et al., "An Antidepressant Decreases CSF Aβ Production in Healthy Individuals and in Transgenic AD Mice," *Science Translational Medicine* 6 (2014): 236re4.

10 Dale E. Bredesen, "Reversal of Cognitive Decline: A Novel Therapeutic Program," *Aging* 6, no. 9 (2014): 707–17.

Chapter 15: Men, Women and Alzheimer's

1 Alzheimer's Association, *2014 Alzheimer's Disease Facts and Figures*, 17.

2 Ibid.

3 A.N.V. Ruigrok et al., "A Meta-Analysis of Sex Differences in Human Brain Structure," *Neuroscience and Biobehavioral Reviews* 39 (February 2014): 34–50.

4 M. Ingalhalikar et al., "Sex Differences in the Structural Connectome of the Human Brain," *Proceedings of the National Academy of Sciences* (early edition), www.pnas.org/cgi/doi/10.1073/pnas.1316909110.

5 C. Fine, "New Insights into Gendered Brain Wiring, or a Perfect Case Study in Neurosexism?" *theconversation.com* (December 4, 2013).

6 M. Spampinato et al., "Gender Differences in Gray Matter Atrophy Patterns in the Progression from Mild Cognitive Impairment to Alzheimer's Disease" (paper presented at the Radiological Society of North America meeting, McCormick Place, Chicago, November 26, 2012).

7 K. Irvine et al., "Greater Cognitive Deterioration in Women Than Men with

Alzheimer's Disease: A Meta Analysis," *Journal of Clinical and Experimental Neuropsychology* 34, no. 9 (November 2012): 989–98.

8 H. Fillit et al., "Observations in a Preliminary Open Trial of Estradiol Therapy for Senile Dementia—Alzheimer's Type," *Psychoneuroendocrinology* 11, no. 3 (1986): 337–45.

9 M. Tang et al., "Effect of Oestrogen during Menopause on Risk and Age at Onset of Alzheimer's Disease," *Lancet* 348 (1996): 429–32; A. Paginini-Hill and V.W. Henderson, "Estrogen Replacement Therapy and Risk of Alzheimer Disease," *Archives of Internal Medicine* 156 (1996): 2213–17; C. Kawas et al., "A Prospective Study of Estrogen Replacement Therapy and the Risk of Developing Alzheimer's Disease," *Neurology* 48 (1997): 1517–21; B. Sherwin, "Can Estrogen Keep You Smart? Evidence from Clinical Studies," *Journal of Psychiatry and Neuroscience* 24, no. 4 (1999): 315–21.

10 S. Shumaker et al., "Conjugated Equine Estrogens and Incidence of Probable Dementia and Mild Cognitive Impairment in Postmenopausal Women," *Journal of the American Medical Association* 291, no. 24 (2004): 2947–58.

11 Ibid., 2947.

12 S.M. Resnick and V.W. Henderson, "Hormone Therapy and Risk of Alzheimer Disease: A Critical Time," *Journal of the American Medical Association* 288, no. 17 (2002): 2170–72.

13 http://www.nih.gov/news/health/jun2013/nia-24.htm.

CHAPTER 16: WAS IT REALLY THE ALUMINUM

1 I. Klatzo et al., "Experimental Production of Neurofibrillary Degeneration," *Journal of Neuropathology and Experimental Neurology* 24, no. 2 (1965): 187–99.

2 Robert Katzman and Katherine Bick, eds., *Alzheimer Disease: The Changing View* (San Diego: Academic Press, 2000), 133.

3 D.R. Crapper et al., "Aluminium, Neurofibrillary Degeneration and Alzheimer's Disease," *Brain* 99 (1976): 67–80.

4 D.R. Crapper et al., "Brain Aluminum Distribution in Alzheimer's Disease and Experimental Neurofibrillary Degeneration," *Science* 180, no. 4085 (1973): 511–13.

5 D.P. Perl and A.R. Brody, "Alzheimer's Disease: X-ray Spectrographic Evidence of Aluminum Accumulation in Neurofibrillary Tangle-Bearing Neurons," *Science* 208 (1980): 297–99.

6 S.E. Levick, "Dementia from Aluminum Pots?" *New England Journal of Medicine* 303 (1980): 164.

7 J.M. Candy et al., "Aluminosilicates and Senile Plaque Formation in Alzheimer's Disease," *Lancet* (February 15, 1986): 354–56.

8 C.N. Martyn et al., "Geographical Relation between Alzheimer's Disease and Aluminium in Drinking Water," *Lancet* 1 (8629) (January 14, 1989): 59–62.

9 A.B. Graves et al., "The Association between Aluminum-Containing Products and Alzheimer's Disease," *Journal of Clinical Epidemiology* 43, no. 1 (1990): 35–44.

10 D.R. Crapper McLachlan et al., "Would Decreased Aluminum Ingestion Reduce the Incidence of Alzheimer's Disease?" *Canadian Medical Association Journal* 145, no. 7 (1991): 793–804.

11 Ibid.

12 J.P. Landsberg et al., "Absence of Aluminium in Neuritic Plaque Cores in Alzheimer's Disease," *Nature* 360 (November 5, 1992): 65–68.

13 Gina Kolata, "New Alzheimer's Study Questions Link to Metal," *New York Times*, November 10, 1992.

14 P. Good and D. Perl, "Aluminium in Alzheimer's?" *Nature* 362 (April 1, 1993): 418.

15 Marvin Ross, "Many Questions But No Clear Answers on Link between Aluminum, Alzheimer's Disease," *Canadian Medical Association Journal* 150, no. 1 (1994): 68–69.

16 David G. Munoz, "Aluminum and Alzheimer's Disease," *Canadian Medical Association Journal* 151, no. 3 (1994): 268.

17 Ibid., 269.

18 Francis Crick, *The Astonishing Hypothesis: The Scientific Search for the Soul* (New York: Touchstone, 1995), xiii.

19 C.S. Martyn et al., "Aluminum Concentrations in Drinking Water and Risk of Alzheimer's Disease," *Epidemiology* 8 (1997): 281–86.

20 David G. Munoz, "Is Exposure to Aluminum a Risk Factor for the Development of Alzheimer Disease?—No," *Archives of Neurology* 55 (May 1998): 737–39.

21 W.F. Forbes and G.B. Hill, "Is Exposure to Aluminum a Risk Factor for the Development of Alzheimer Disease?—Yes," *Archives of Neurology* 55 (May 1998): 740–41.

22 C. Exley and M.M. Esiri, "Severe Cerebral Congophilic Angiopathy Coincident with Increased Brain Aluminium in a Resident of Camelford, Cornwall, UK," *Journal of Neurology, Neurosurgery and Psychiatry* 77 (2006): 877–79.

23 A. David and S. Wessely, "The Legend of Camelford: Medical Consequences of a Water Pollution Accident," *Journal of Psychosomatic Research* 39, no. 1 (1995): 1–9.

24 L. Tomljenovic, "Aluminum and Alzheimer's Disease: After a Century of Controversy, Is There a Plausible Link?" *Journal of Alzheimer's Disease* 23 (2011): 567–98.

25 Ibid., 577.

26 M. Kawahara and M. Kato-Negishi, "Link between Aluminum and the Pathogenesis of Alzheimer's Disease: The Integration of the Aluminum and Amyloid Cascade Hypotheses," *International Journal of Alzheimer's Disease*, Article ID 276393 (2011): 1–17.

CHAPTER 17: THE MANY FACES OF DEMENTIA

1 E.E. Manuelidis and L. Manuelidis, "Suggested Links between Different Types of Dementias: Creutzfeldt-Jakob Disease, Alzheimer Disease, and Retroviral CNS Infections," *Alzheimer Disease and Associated Disorders* 3 (1989): 100–109.

2 D.J. Irwin et al., "Evaluation of Potential Infectivity of Alzheimer and Parkinson Disease Proteins in Recipients of Cadaver-Derived Human Growth Hormone," *JAMA Neurology* 70, no. 4 (April 2013): 462–68.

3 In Richard Rhodes, *Deadly Feasts* (New York: Simon and Schuster, 1997), 143.

4 http://www.mvguam.com/local/news/22139-neurologist-neurodegenerative-disease-may-end-on-guam.html#.U9PeEkhFHFw.

5 D.P. Perl et al., "Calculation of Intracellular Aluminum Concentration in Neurofibrillary Tangle (NFT) Bearing and NFT-Free Hippocampal Neurons of ALS/Parkinsonism-Dementia of Guam Using Laser Microprobe Mass Analysis (LAMMA)," *Journal of Neuropathology and Experimental Neurology* 45 (1986): 379.

6 M.G. Whiting, "Toxicity of Cycads," *Economic Botany* 17 (1963): 271–302.

7 P.A. Cox and O.W. Sacks, "Cycad Neurotoxins, Consumption of Flying Foxes and the ALS-PDC Disease of Guam," *Neurology* 58 (2002): 956–59.

8 S.A. Banack and P.A. Cox, "Biomagnification of Cycad Neurotoxins in Flying Foxes," *Neurology* 61 (2003): 387–89.

9 A.R. Borenstein et al., "Cycad Exposure and Risk of Dementia, MCI and PDC in the Chamorro Population of Guam," *Neurology* 68 (2007): 1764–71.

10 E. Masseret et al., "Dietary BMAA Exposure in an Amyotrophic Lateral Sclerosis Cluster from Southern France," *PLoS One* 8, no. 12 (December 2013): e83406.

11 Jason Richardson et al., "Elevated Serum Pesticide Levels and Risk for Alzheimer Disease," *JAMA Neurology* 71, no. 3 (March 2014): 284–90.

Chapter 18: Where You Live, What You Eat

1 http://www.braininstitute.ca/sites/default/files/final_report_obi_pa_alzheimers_february_25_2013.pdf.

2 J.A. Duke, "Turmeric, the Queen of Cox-2-Inhibitors," *Alternative and Complementary Therapies* (October 2007): 229–34.

3 F. Yang et al., "Curcumin Inhibits Formation of Amyloid Oligomers and Fibrils, Binds Plaques, and Reduces Amyloid *in Vivo*," *Journal of Biological Chemistry* 280, no. 7 (2005): 5892–5901.

4 S.A. Frautschy, "Phenolic Anti-inflammatory Antioxidant Reversal of Ab-induced Cognitive Deficits and Neuropathology," *Neurobiology of Aging* 22 (2001): 993–1005.

5 N. Hishikawa, "Effects of Turmeric on Alzheimer's Disease with Behavioral and Psychological Symptoms of Dementia," *Ayu* 33, no. 4 (October–December 2012): 499–504.

6 L. Baum et al., "Six-Month Randomized, Placebo-Controlled, Double-Blind, Pilot Clinical Trial of Curcumin in Patients with Alzheimer Disease," *Journal of Clinical Psychopharmacology* 28, no. 1 (February 2008): 110–12.

7 J.M. Ringman et al., "Oral Curcumin for Alzheimer's Disease: Tolerability and Efficacy in a 24-Week Randomized, Double Blind, Placebo-Controlled Study," *Alzheimer's Research and Therapy* 4, no. 43 (2012): 1–8.

8 T. Ng et al., "Curry Consumption and Cognitive Function in the Elderly," *American Journal of Epidemiology* 164, no. 9 (July 26, 2006): 898–906.

9 V. Chandra et al., "Incidence of Alzheimer's Disease in a Rural Community in India," *Neurology* 57 (2001): 985–89.

10 V. Patel and M. Prince, "Ageing and Mental Health in a Developing Country: Who Cares? Qualitative Studies from Goa, India," *Psychological Medicine* 31 (2001): 29–38.

11 Dr. Mary Ganguli to author (email), April 22, 2014.

12 H.C. Hendrie et al., "Incidence of Dementia and Alzheimer Disease in 2 Communities," *Journal of the American Medical Association* 285, no. 6 (February 14, 2001): 739–47.

13 L. White et al., "Prevalence of Dementia in Older Japanese-American Men in Hawaii," *Journal of the American Medical Association* 276, no. 12 (1996): 955–60.

14 M.C. Morris et al., "Associations of Vegetable and Fruit Consumption with Age-Related Cognitive Change," *Neurology* 67 (2006): 1370–76.

15 P.K. Crane et al., "Glucose Levels and Risk of Dementia," *New England Journal of Medicine* 369, no. 6 (August 8, 2013): 540–48.

16 S. de la Monte, "Brain Insulin Resistance and Deficiency as Therapeutic Targets in Alzheimer's Disease," *Current Alzheimer Research* 9 (2012): 35–66.

17 J. Mehla et al., "Experimental Induction of Type 2 Diabetes in Aging-Accelerated Mice Triggered Alzheimer-Like Pathology and Memory Deficits," *Journal of Alzheimer's Disease* 39 (2014): 145–62.

18 S. de la Monte and J.R. Wands, "Alzheimer's Disease Is Type 3 Diabetes: Evidence Reviewed," *Journal of Diabetes Science and Technology* 2, no. 6 (November 2008): 1101–13.

19 M.A. Reger et al., "Intranasal Insulin Administration Dose-Dependently Modulates Verbal Memory and Plasma Amyloid-β in Memory-Impaired Older Adults," *Journal of Alzheimer's Disease* 13 (2008): 323–31.

20 S. de la Monte, "Brain Insulin Resistance," 42.

CHAPTER 19: WHAT'S NEXT?

1 J.W. Vaupel and H. Lundström, in *Studies in the Economics of Aging,* ed. D.A. Wise (Chicago: University of Chicago Press, 1994): 79–104.

2 K.C. Manton et al., "Declining Prevalence of Dementia in the U.S. Elderly Population," *Advances in Gerontology* 16 (2005): 30–37; E.M.C. Schrijvers et al., "Is Dementia Incidence Declining? Trends in Dementia Incidence since 1990 in the Rotterdam Study," *Neurology* 78 (May 2012): 1456–63; F.E. Matthews et al., "A Two-Decade Comparison of Prevalence of Dementia in Individuals Aged 65 Years and Older from Three Geographical Areas of England: Results of the Cognitive Function and Ageing Study I and II," *Lancet* 382, no. 9902 (October 26, 2013): 1405–12.

3 M. Mapstone et al., "Plasma Phospholipids Identify Antecedent Memory Impairment in Older Adults," *Nature Medicine* 20 (2014): 415–18; A. Hye et al., "Plasma Proteins Predict Conversion to Dementia from Prodromal Disease," *Alzheimer's and Dementia* (2014): 1–9.

Index